WOOD FRAME HOUSEBUILDING

AN ILLUSTRATED GUIDE

To John Wahlfeldt

Popular Science Books offers a wood identification kit that includes 30 samples of cabinet woods. For details on ordering, please write: Popular Science Books, P.O. Box 2033, Lathan, N.Y. 12111.

No. 3005
$21.95

WOOD FRAME HOUSEBUILDING

AN ILLUSTRATED GUIDE

**Bette
Galman
Wahlfeldt**

TAB BOOKS Inc.
Blue Ridge Summit, PA

FIRST EDITION
FIRST PRINTING

Library of Congress Cataloging in Publication Data

Wahlfeldt, Bette G. (Bette Galman)
Wood frame housebuilding, an illustrated guide / by Bette Galman
Wahlfeldt.
p.   cm.
Includes index.
ISBN 0-8306-0405-7   ISBN 0-8306-9305-X (pbk.)
1. Wooden-frame houses.   2. House construction.   I. Title.
TH4818.W6W34   1988   88-1893
694—dc 19   CIP

Questions regarding the content of this book
should be addressed to:

Reader Inquiry Branch
TAB BOOKS Inc.
Blue Ridge Summit, PA 17294-0214

Contents

Acknowledgments

It is with deep appreciation that I thank the following people for their help in providing information for this work: Maryann Ezell, American Plywood Association; Raymond W. Moholt, Western Wood Products Association; Donald D. Rowell, Leslie-Locke, A Questor Company; Larry Kelleher, All Nighter Stove Works, Inc.; I. Melody, Florida Solar Energy Center; Clem H. Moreman, Birmingham Stove & Range Co.; Murray Feiss, Murray Feiss Import Corp; Gregory R. Olsen, Celotex Corp.; Patricia Logan, Costich & McConnell; Stacey G. Wilson, Georgia-Pacific Corp.; Donald W. Minton, Keller Crescent Co.; Mary Sue Baron, Bell Eanes Advertising for Morton Miller, General Products Co.; Edgar Poe, United States Dept. of Agriculture. Without their unselfish sharing, I would have been hard-pressed to get my work completed.

Introduction

"Let's get back to basics." This phrase has become part of the American vernacular in the past decade. It refers to a renewed interest in the lifestyle of our forefathers, everything from education and food, to housebuilding.

With interest rates and the costs of building at an all-time high, we have to find ways of getting more dollar value for less cash outlay if we are to continue to build good, sound, quality housing. Many people are accomplishing this by doing much or all of the work involved in building a house themselves.

Perhaps you are wondering where to begin. Ask yourself these basic questions: How big a house do you need? How much money do you have to spend? Will it be a permanent home that will be resold if your family outgrows it, or do you want a starter home that you can add to as your family expands? And, of course, how much do you know about building? Don't let the large amount of questions and uncertainties overwhelm you. Take them one at a time.

If you are an experienced handyman, you already know there are many projects within your abilities. However, even if you're a beginner, by studying the "basics" you'll find that there are many phases of building that, with a little patience, even the inexperienced can tackle.

Start by making a list of the jobs you know you can handle. Then, make a list of those you've never handled but are willing to expend time and effort in trying.

The following chapters take you on a step-by-step tour and guide you through the various aspects of building your home. Remember to make a checklist of the equipment you'll need for all phases of construction. Even if you build your entire house—and many people can and do—you don't need a million-dollar tool and equipment inventory. For basic building you'll be using basic tools: hammers, saws, screwdrivers, tape measures, leveling measures, hand saws. Even for those large jobs such as pouring concrete, no great cash outlay is required if you make use of the handy rental stores found in most cities.

From pouring concrete footings to erecting a fence, from insulating to landscaping, finishing the interior to painting outside, installing a fireplace to building a garden/patio deck, you'll be amazed and delightfully surprised to find there is much you and your family can do in the course of building your home. For those who cannot handle the entire building project for whatever reason, choose the projects you feel comfortable doing and subcontract the rest.

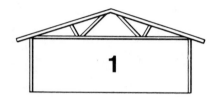

Overview

Choosing lumber was a less confusing task for our parents and grandparents than it is for us today. Locally grown species were usually used and grades and species selections were minimal. Today there are a number of softwood species from which to choose, and dimension lumber grades have been developed on a more precise engineering basis.

There are a number of good reasons to use wood siding on houses. One is that it will keep the bright new "face" of any house attractive for many years. In fact, given reasonable care, wood siding will retain its beauty for centuries, as witnessed by some skill-beautiful siding on houses that date back to early colonial times.

Using wood and wood-based materials for siding makes sense for design as well as structural reasons. Its great flexibility and variety in patterns, sizes, and colors range from the captivating simplicity of colonial clapboard to the equally charming board-and-batten and rough-textured plywood or lumber siding of modern ranch and rambler houses. In between are many other distinctive and attractive varieties and combinations.

Cost is yet another good reason for using for wood siding. Wood has a relatively low initial cost and can be sawed, fitted, and nailed into place with ease. It is fuel-saving, due to the millions of tiny hollow fibers that make it a good insulator, and it has natural resistance to the ravages of weather.

Other benefits include wood's ability to hold a wide variety of finishes—clear finishes that reveal and accentuate its natural beauty, stains that give a rustic appearance, and paints of every conceivable color. Within limits, the color scheme can be changed periodically to give a home a fresh new appearance. It is little wonder that wood siding has come to stand for warmth, comfort, security, beauty—all the things that spell "home."

There are many kinds of wood siding on the market today (Fig. 1-1). All will give good service, provided each is used correctly. For houses, bevel siding is perhaps the most widely used. Drop siding and shiplap patterns are also used, especially on houses without sheathing. These patterns of siding are applied horizontally and tend to make a house appear lower and longer.

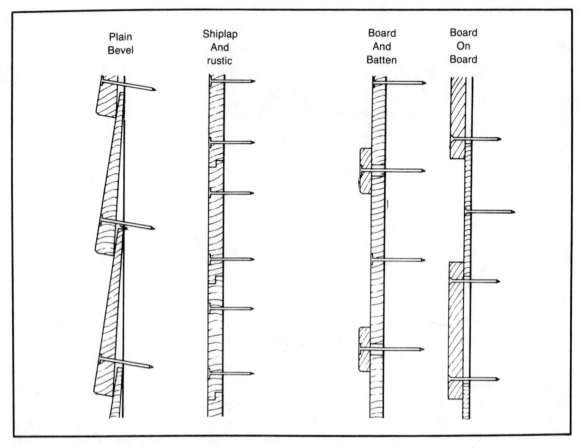

| Plain Bevel | Shiplap And rustic | Board And Batten | Board On Board |

Fig. 1-1. Patterns of wood siding. Courtesy USDA

Vertical siding is increasingly popular on one-story houses. It consists of tongue-and-groove boards (Fig. 1-2) or square-edge boards applied vertically, usually with narrow strips called battens nailed over the joints. Vertical patterns with smooth and rough-sawn surfaces are available in plywood or Douglas fir, redwood, and western red cedar. Vertical siding tends to make a house appear taller.

The kind of wood siding to use depends on where the house is built, its price range, and the desired architectural effect. You will find a wide quality range in lumber—from that with a clear, smooth grain to that with a rough, flat grain containing knots and other characteristics that should be given special finishing treatments.

Before you begin to build your wood frame structure(s) learn as much as you can about the spe-

cies (types) of wood available in your area so you can make the best choice for your particular situation. Visit the local lumberyard to find out what is available in your locality, and become familiar with the most commonly marketed species and grades in your area. Wood will also be differentiated by the extent to which it has been manufactured—that is,

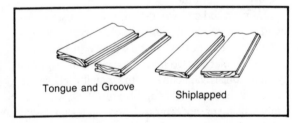

Tongue and Groove   Shiplapped

Fig. 1-2. Tongue-and-groove and shiplapped lumber. Courtesy USDA

rough-sawn, dressed, and worked, tongue-and-groove, shiplapped, or patterned.

LUMBER GRADES

Proper and uniform grading and grade marking is essential to comply with building codes. Grading also assures the buyer of obtaining pieces in each grade with the same range of strength and appearance properties regardless of manufacturer's species or log quality. Grading provides the buyer with a dependable yardstick for determining relative lumber values in the market and facilitates specifying and checking.

Grading of lumber is based upon a visual inspection and is primarily a judgment of end-use strength rather than appearance. Natural log characteristics that have an effect upon strength are taken into account along with manufacturing imperfections.

Illustrations are used when grading that are representative of grades of western softwood dimension lumber. Grades are established by the National Grading Rule Committee, acting through the American Lumber Standards Committee under procedures set up by the U.S. Department of Commerce. Each illustration includes examples of allowable characteristics in each grade.

Grading practices of the mills who belong to the Western Wood Products Association (WWPA) are closely supervised by this Association to assure uniformity. The resulting grades provide a dependable measure for determining the value of lumber.

The many mills manufacturing lumber from the same or similar woods apply the grade stamp to indicate that they employ stringent quality control and standards to achieve a better than 95-percent probability that an individual piece of lumber will equal the predicted average strengths for the grade.

The official WWPA grade stamp on a piece of lumber is assurance of its assigned grade. Grade stamps are intended to signify that the lumber conforms to the the standards in the rule books, subject to the reasonable variation between graders, and is based on visual inspection. These stamps should not be taken as affirmation that each stamped piece of lumber is defect-free.

Before you purchase a large amount of lumber for your construction it is a good idea to learn a little about lumber grades. The symbols in Fig. 1-3 may appear in various combinations on official WWPA gradestamps.

Figure 1-3(A) is the official Association certification mark.

Figure 1-3(B) indicates the mill. Each mill is assigned a permanent number. Some mills are identified by mill name or abbreviation instead of by mill number Figure 1-3(C) is an example of an official grade name abbreviation. In this case 2 COM stands for 2 Common boards as described in the WWPA 1970 Grading Rules. Its appearance in grade mark identifies the grade of a piece of lumber.

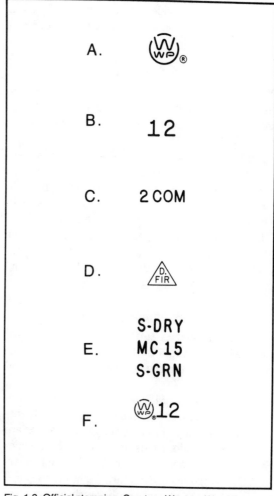

Fig. 1-3. Official stamping. Courtesy Western Wood Products

Figure 1-3(D) denotes a species mark identifying the free species from which the lumber is sawn, in this case, Douglas fir.

The marks in Fig. 1-3(E) denote the moisture content of the lumber when manufactured. S-DRY indicates a moisture content not exceeding 19 percent. MC 15 indicates a moisture content not exceeding 15 percent. S-GRN indicates that the moisture content exceeded 19 percent.

When an inspection certificate issued by the Western Wood Products Association is required on a shipment of lumber and specific grade marks are not used, the stock is identified by an imprint of the Association and the number of the shipping mill or inspector. Figure 1-3(F) is an example.

Figure 1-4 provides some samples of various grades that appear on wood purchased at your local lumberyard.

Species

Western softwood species commonly manufactured into dimension lumber include western hemlock, Engelmann spruce, larch, western cedars and all pines and firs.

Many of these species are grown, harvested, manufactured, and marketed together, and have similar performance properties which make them interchangeable in use, grading, and grade marking. Douglas fir and larch are grouped together as "Douglas-Fir-larch." Western hemlock combines with true firs as "hem-fir." The "white woods" include Engelmann spruce, all true firs, hemlocks, and pines. Ponderosa pine is an individual species.

Specifications

Dimension lumber is surfaced lumber of nominal thickness from 2 to 4 inches. It is used for structural support and framing, including studs, joists, and rafters.

National Grading Rules classify dimension lumber in three width categories and four use categories. Structural Light Framing and Light Framing are 2 to 4 inches wide. Studs are 2 to 6 inches wide. Structural Joists and Planks are 5 inches and wider.

The categories of Structural Light Framing and Structural Joists and planks each contain four grades,

indicating a range of allowable characteristics and manufacturing imperfections affecting strength, stiffness, and appearance.

Select Structural is the highest grade in Structural Light Framing and Structural Joists and Planks. Such wood recommended where good appearance is required, along with strength and stiffness.

No. 1 grade is recommended where good appearance is desired but is secondary to strength and stiffness. No. 2 grade is recommended for most general construction uses. No. 3 grade is appropriate for general construction where high strength is not necessary.

The Light Framing category contains three grades. Construction is the highest grade in Light Framing, indicating a piece of wood widely used for general framing, with good appearance but graded primarily for strength and serviceability. Some pieces in this grade would be No. 1 or better in Structural Light Framing.

Standard grade is customarily used for the same purposes, or in conjunction with construction grade, providing good strength and excellent serviceability.

Utility grade is recommended for studding, blocking, plates and bracing—where economy is desired.

Stud grade is a separate grade, suitable for all stud uses including load-bearing walls. This is one of the most popular grades for wall construction, although lengths are limited to 10 feet.

Economy grades are also available in all three categories of Structural Light Framing, Light Framing, and Structural Joists and Planks.

Characteristics

Grades are determined primarily by the natural characteristics of the log which appear in a given piece of lumber and which have an effect upon its strength, stiffness, and appearance. Manufacturing imperfections, no matter what the cause, also affect the grade.

Knots are the most frequently encountered characteristics. There are various types of knots. In Figs. 1-5 and 1-6, the most common types as they appear in the bottom photo shows the knots on the lumber face in cross-section. In Fig. 1-5 the knots

Fig. 1-4. Grade marks. Courtesy Western Wood Products

are: (A) round knot hole through two side faces, resulting from loose knot; (B) Sound encased, fixed, round knot through two wide faces; and (C) sound, watertight, tight, intergrown knot with radial checks. In Fig. 1-6 the knots are: (A) sound, tight, round pitch knot through two narrow faces; (B) intergrown round knot through all four faces; and (C) sound, tight, intergrown, watertight spike knot through three faces.

Wane is the presence of bark or lack of wood from any cause on the edge or corner of a piece of lumber (Fig. 1-7).

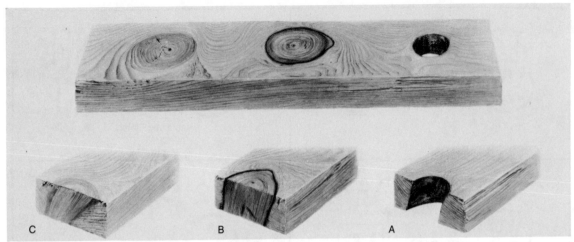

Fig. 1-5. Knots. Courtesy Western Wood Products

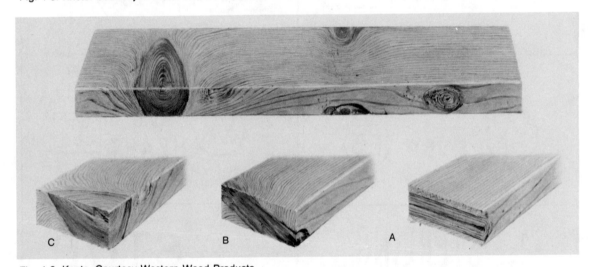

Fig. 1-6. Knots. Courtesy Western Wood Products

Fig. 1-7. Wave. Courtesy Western Wood Products

6

Twist is a deviation from the flat planes of all four faces by a spiraling or torsional action, usually the result of seasoning (Fig. 1-8).

Bow is a deviation from a flat plane of the wide face of a piece of lumber from end to end (Fig. 1-9).

Crook is a deviation from a flat plane of the narrow face of lumber from end to end (Fig. 1-10).

Here are some other characteristics to look for: Shake is a lengthwise separation of the wood that usually occurs between or through the annual growth rings. White speck and honeycomb are caused by a fungus in the living tree. White speck appears as small white pits or spots. Honeycomb is similar but the pits are deeper or larger. Neither is subject to further decay unless used under wet conditions.

Decay is disintegration of the wood due to action of wood-destroying fungi. Checks and separations of the wood fibers, normally occurring across or through the annual growth rings, are usually as a result of seasoning. They may occur anywhere on a piece.

Splits are similar to checks except the separa-

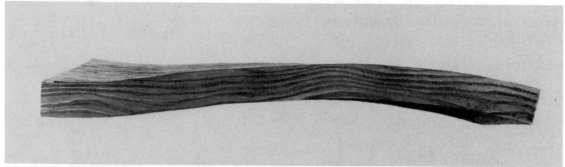

Fig. 1-8. Twist. Courtesy Western Wood Products

Fig. 1-9. Bow. Courtesy Western Wood Products

Fig. 1-10. Crook. Courtesy Western Wood Products

tions of the wood fibers extend completely through a piece, usually at the ends. Cup is a deviation from a flat plane, edge to edge.

Additional characteristics include, burl, compression wood, pitch streak, pith, picket, sapwood, and stain.

Manufacturing imperfections include chipped, torn, raised, or loosened grain; skips in surfacing; mismatch wavy dressing; and machine-caused burns, chips, and bite or knife marks.

You'll find examples of all of these at your local lumberyard. Pay a visit and discuss the lumber you plan to purchase well in advance to make sure you get just the kind of lumber that is right for your building.

TYPES OF WOOD

To assist you in recognizing the differences in woods, a brief explanation of each follows.

Ponderosa Pine

In Select Structural grades, knots are limited to sound, firm, encased, and pith knots, if tight and well-spaced, with only one unsound or loose knot or hole per 4 lineal feet. Centerline knots range from maximums of 1½-on-5-inch widths to 3¼-on-14-inch widths. Edge knots range from maximums of 1-on-5-inch widths to 2½-on-14-inch widths. Unsound or loose knots or holes range from maximums of ⅞-on-4-inch widths to 1½-on-14-inch widths, one per 4 lineal feet. (Figs. 1-11 and 1-12)

Fig. 1-11. Ponderosa pine. Courtesy Western Wood Products

Fig. 1-12. Ponderosa pine, structural select. Courtesy Western Wood Products

Hem-Fir

In Select Structural grades, knots are limited to sound, firm, encased, and pitch knots, if tight and well-spaced, with only one unsound or loose knot or hole per 4 lineal feet. Centerline knots range from maximums of 1½-on-5-inch widths, to 3¼-on-14-inch widths. Edge knots range from maximums of 1-on-5-inch widths to 2³⁄₈-on-14-inch widths. Unsound or loose knots or holes range from maximums of ⅞-on-5-inch widths to 1¼-on-14-inch widths, one per 4 lineal feet (Fig. 1-13).

In Structural JP No. 1, knots must be of the same type as in Select Structural grade but can be slightly larger, with one unsound or loose knot or hole permitted per 3 lineal feet (Fig. 1-14). In No.

2, well-spaced knots of any quality are allowable with one hole from any cause per 2 lineal feet (Fig. 1-15). In No. 3, well-spaced knots of any quality are allowable, with only one hole from any cause per lineal foot (Fig. 1-16).

Select Structural Douglas fir-larch has knots that are limited to sound, firm, encased, and pitch knots, if tight and well-spaced, with one unsound or loose knot or hole per 4 lineal feet. Centerline knots range from maximums of 1-on-5-inch widths to 2⅜-on-14-inch widths. Unsound or loose knots or holes range from maximums of ⅛-on-5-inch widths to 1¼-on-14-inch widths, one per lineal foot.

In Select Structural Douglas fir-larch No. 1, knots must be of the same type as in Select Struc-

9

Fig. 1-13. Hem-fir structural JP. Courtesy Western Wood Products

tural grade but can be slightly larger, with only one unsound or loose knot or hole permitted per 3 lineal feet.

Characteristics seen in Light Framing (hem-fir) construction grade are sound, firm, encased, and pitch knots that are tight and no larger than 1½ inch. Unsound or loose knots or holes are limited to 1 inch, one per 3 lineal feet.

Light Framing Douglas fir-larch stud knots are not restricted as to quality but must be well-spaced and of sizes up to what is allowable in Utility grade. This separate grade, suitable for all stud uses including load-bearing walls, places limitations on crook, wane, and edge knots. Lengths are limited to a maximum of 10 feet.

Utility knots are not restricted as to quality up to 2½ inches anywhere on the wide face. Holes are limited to 1½ inch, one per lineal foot.

MATERIALS SELECTION AND USE

Wood in its various forms is the most common material used in house construction. It is used for framing of floors, walls, and roofs, It is sometimes used in board form as a covering material, but more often the covering materials are plywood or other panel wood products. Wood is also used as siding or exterior covering, as interior covering, as interior and exterior trim, as flooring, in the many forms and types of millwork, and as shingles to cover roofs and

10

sidewalls (although the newer fiberglass shingles are recommended for this purpose).

Wood is easy to form, saw, nail, and fit—even with simple handtools. With proper use and protection it will give excellent service. The moisture content of wood used in various parts of a house is important, and recommended moisture contents will be outlined in later sections.

There are a number of basic standard wood and wood products used in the construction of wood-frame houses. The materials can be grouped according to their use in the construction of a house. Some are important for good strength, others for workability, and still others for good appearance.

Treated Posts

Wood posts or poles that are embedded in the soil and used for support of the house should be pressure treated. A number of species are used for these round members. The pressure treatments should conform to Federal Specifications TT-W-571. Pressure treatments normally utilize the oil type of preservatives (empty-cell process) or leach-resistant waterborne salt preservatives (full-cell process).

Dimension Material

Surfaced dimension material wood members as you purchase them from the lumber yard are not the size

Fig. 1-14. Hem-fir No. 1. Courtesy Western Wood Products

Fig. 1-15. Hem-fir No. 2. Courtesy Western Wood Products

they originally were. For example, a nominal 2-by-4 might have a finished thickness of 1½ to 1⁹⁄₁₆ inches and a width of 3½ to 3⁹⁄₁₆ inches, depending on the moisture content. These materials are sawn from green logs and must be surfaced and dried. These processes account for the difference in size between a finished dry member and a rough green member. Table 1-1 provides a breakdown of the nominal and minimum-dressed sizes of finished boards, dimension, and timbers. The thicknesses apply to all widths, and all widths to all thicknesses.

The first materials to be used, after the foundation is in place, are the floor joists and the beams upon which the joists rest. These require adequate strength in bending and moderate stiffness. A spe-

cies, such as southern pine, western hemlock, or Douglas fir is commonly selected for these uses. In lower-cost houses, the third grade is usually acceptable.

For best performance, the moisture content of most dimension materials should not exceed 19 percent.

Wall studs (the structural members making up the wall framing) are usually nominal 2 by 4 inches in size and spaced 16 or 24 inches apart. The strength and stiffness of these are not as important as that of the floor joists, and in low-cost houses the third grade of a species such as Douglas fir or southern pine is satisfactory. Slightly higher grades of other species, such as white fir, eastern white pine, spruce, the western white pines, and others, are normally used.

Members used for trusses, rafters, beams, and ceiling joists in the roof framing have about the same requirements as those listed for floor joists. For low-cost houses, the second grade can be used for trusses if the additional strength reduces the amount of material required.

Covering Materials

Floor sheathing (subfloor) consists of board lumber or plywood, and the intended use determines thickness of the subfloor. A single layer may serve both as subfloor and top surface material; for example, ⅝-inch or thicker tongue-and-groove plywood of Douglas fir, southern pine, or other species in a slightly greater thickness can be used when joints are spaced no more than 20 inches on center.

While Douglas fir and southern pine plywoods are perhaps the most common, other species are equally adaptable for floor, wall, and roof coverings. The Identification Index system of marking each sheet of plywood provides the allowable rafter or roof truss and floor joist spacing for each thickness of a standard grade suitable for this purpose. A nominal 1-inch board subfloor normally requires a top covering of some type.

Roof sheathing, like the subfloor, most commonly consists of plywood or board lumber. Where exposed wood beams spaced 2 to 4 feet apart are

Fig. 1-16. Hem-fir No. 3. Courtesy Western Wood Products

Table 1-1. Nominal and Minimum-Dressed Sizes of Finished Boards, Dimension, and Timbers.

ITEM	THICKNESSES				FACE WIDTHS			
	NOMINAL	MINIMUM DRESSED			NOMINAL	MINIMUM DRESSED		
		Inches				Inches		
Select or Finish (19 per cent moisture content)	3/8	5/16			2	1 1/2		
	1/2	7/16			3	2 1/2		
	5/8	9/16			4	3 1/2		
	3/4	5/8			5	4 1/2		
	1	3/4			6	5 1/2		
	1 1/4	1			7	6 1/2		
	1 1/2	1 1/4			8	7 1/4		
	1 3/4	1 3/8			9	8 1/4		
	2	1 1/2			10	9 1/4		
	2 1/2	2			11	10 1/4		
	3	2 1/2			12	11 1/4		
	3 1/2	3			14	13 1/4		
	4	3 1/2			16	15 1/4		
		Dry Inches	Green Inches			Dry Inches	Green Inches	
Boards	1	3/4	25/32		2	1 1/2	1 9/16	
	1 1/4	1	1 1/32		3	2 1/2	2 9/16	
	1 1/2	1 1/4	1 9/32		4	3 1/2	3 9/16	
					5	4 1/2	4 5/8	
					6	5 1/2	5 5/8	
					7	6 1/2	6 5/8	
					8	7 1/4	7 1/2	
					9	8 1/4	8 1/2	
					10	9 1/4	9 1/2	
					11	10 1/4	10 1/2	
					12	11 1/4	11 1/2	
					14	13 1/4	13 1/2	
					16	15 1/4	15 1/2	
Dimension	2	1 1/2	1 9/16		2	1 1/2	1 9/16	
	2 1/2	2	2 1/16		3	2 1/2	2 9/16	
	3	2 1/2	2 9/16		4	3 1/2	3 9/16	
	3 1/2	3	3 1/16		5	4 1/2	4 5/8	
					6	5 1/2	5 5/8	
					8	7 1/4	7 1/2	
					10	9 1/4	9 1/2	
					12	11 1/4	11 1/2	
					14	13 1/4	13 1/2	
					16	15 1/4	15 1/2	
					2	1 1/2	1 9/16	
					3	2 1/2	2 9/16	
					4	3 1/2	3 9/16	
					5	4 1/2	4 5/8	
Dimension	4	3 1/2	3 9/16		6	5 1/2	5 5/8	
	4 1/2	4	4 1/16		8	7 1/1	7 1/2	
					10	9 1/4	9 1/2	
					12	11 1/4	11 1/2	
					14		13 1/2	
					16		15 1/2	
Timbers	5 & Thicker	1/2 Off			5 & Wider	1/2 Off		

used for low-pitched roofs, for example, wood decking, fiberboard roof deck, or composition materials in 1-to-3-inch thicknesses might be used. The thickness varies with the spacing of the supporting rafters or beams. These sheathing materials often serve as an interior finish as well as a base for roofing.

Wall sheathing, if used with a siding of secondary covering material, can consist of plywood, lumber, structural insulating board, or gypsum board. The type and method of sheathing application normally determines whether corner bracing is required in the wall. When 4-by-8-foot sheets of 25/32-inch regular or 1/2-inch thick, medium-density, insulating fiberboard or 4/16-inch or thicker plywood are used vertically with proper nailing all around the edge, no bracing is required to resist windstorms. There are plywood materials available with grooved or roughened surfaces that serve both as sheathing and finishing materials. Horizontal application of plywood, insulating fiberboard, lumber, and other materials usually requires some type of diagonal brace for rigidity and strength.

Exterior Trim

Some exterior trim, such as facia boards at cornices or gable-end overhangs, is put in place before the roofing is applied. You will save money if you use only those materials necessary for good utility and a satisfactory appearance. These trim materials are usually wood, and if relatively clear of knots, they can be painted without problems. Lower-grade boards with a rough-sawn surface can be stained.

Roofing

One of the lowest cost roofing materials satisfactory for sloped roofs is mineral-surfaced asphalt roll roofing. Asphalt and fiberglass shingles also give good surface. All come in a variety of colors. The material cost of asphalt shingles and fiberglass shingles is higher than that of surfaced roll roofing. Roll roofing is best saved for additions to the house. The asphalt and fiberglass shingles are more expensive, but last longer and have a better appearance than the roll roofing.

Window and Door Frames

Double-hung, casement, or awning wood windows normally consist of a prefitted sash in assembled frames, ready for installation. A double-hung window is one in which the upper and lower sash is hinged at the side and swings in or out. An awning window is hinged at the top and swings out. A separate sash in $1\frac{1}{8}$ or $\frac{1}{38}$-inch thickness can be used, but some type of frame must be made that includes jamb, stops, sill casing, and the necessary hardware. A fixed sash or a large window glass can be fastened by stops to a prepared frame and generally costs less than a moveable-type window.

It is normally more economical to use one larger window unit than two smaller ones. Screens are ordinarily supplied for all operable windows and for doors. In the colder climates storm windows and storm or combination doors are important for energy savings. In more temperate climates they keep the hot air out and the cool air in. Combination units, with screen and storm inserts, are commonly used and are the ultimate in efficiency.

Exterior Coverings

Exterior coverings such as horizontal wood siding, vertical boards, boards and battens, and similar types of siding usually require some type of backing in the form of sheathing or nailers between studs. In mild climates, nominal 1-inch and thicker sidings are often used over a waterproof paper applied directly to the braced stud wall. There are many sidings of this type on the market in both wood and nonwood materials.

Combination sheathing siding materials (panel siding) usually consist of 4-foot-wide sheets of plywood, exterior particleboard, or hardboard. Applied vertically before installation of window and door frames, such materials serve very well for exteriors. Plywood should be stained or painted, and the other materials painted. Paper-overlaid plywood also serves as a dual-purpose exterior covering material and takes paint well. Wood shingles and shakes and similar materials normally require a solid backing or spaced boards of some type.

Insulation

Most houses, even the lowest cost, must have some type of insulation to resist both cold and heat. There are various types of insulation, from insulating fiberboard to fill types, which can be used in the construction of a house. Perhaps the most common thermal insulations are the flexible (blanket and batt) and the fill types.

A blanket insulation can be used between the floor joists or studs. Batt insulation of various types might be used between floor joists or in the ceiling areas. Most flexible insulations are supplied with a vapor barrier that resists movement of water vapor through the wall and minimizes condensation problems. A friction-type batt insulation is also available for use in floors, walls, or ceilings. Fill-type insulation is most commonly used in attic ceiling areas.

The structural insulating board often serves as sheathing in the wall or as a fair insulating material under a plywood floor. Base selection of insulation on climate as well as on cost and utility.

Interior Coverings

Many drywall (unplastered) interior coverings are

available, from gypsum board to prefinished plywood. Perhaps the most economical are the gypsum board products. They are normally applied vertically in 4-by-8-foot sheets or horizontally in room-length sheets with the joint at midwall heights. They are also used for ceilings. Thicknesses range from ⅜ to ⅝ inches. Butt joints and corners require the use of tape and joint compound, or a corner bead, which adds to labor costs over prefinished materials. Plastic-covered gypsum board is also available at additional cost, but must usually be installed with an adhesive.

Hardboards, insulation board, plywood, and other sheet materials are available, as are wood and fiberboard paneling. The choice must be based on overall cost of material and labor as well as on ease of maintenance. Prefinished ceiling tile in 12-by-12-inch sizes and larger can also be used.

Interior Finish and Millwork

Interior finish and millwork consist of doors and door frames, base moldings, window and door trim, kitchen and other cabinets, flooring, and similar items. The type and grade determine the cost. Selecting simple moldings, cheaper species for jambs and other wood members, simple kitchen shelving, low-cost floor coverings, and eliminating doors where practical can save you hundreds of dollars in total cost of the house.

The most commonly used doors are the flush and panel types. The flush type consists of thin plywood or similar facings with a solid or hollow core. The panel door consists of solid side stiles and cross-rails with plywood or other panel fillers. For exterior types, both may be supplied with openings for glass. Exterior doors are usually 1¾ inch thick and interior doors 1⅜ inch thick.

Door jambs, casings, moldings, and similar millwork are made of any number of wood species. Select the lower-cost materials, yet those that will still give good service—pines, spruces, and Douglas-fir, for example.

Wood strip flooring or hardwood wood tile might be too costly to consider in the original construction, but could be installed at a future time. Softwood floorings or the lower cost hardwood floorings are more within the original budget. Asphalt tile or even a painted finish are other cheaper alternatives. However, when a woodboard subfloor is used, an underlayment of particleboard, hardboard, or plywood is required under the tile.

Nails and Nailing

In a wood-frame house, nailing is the most common method of fastening the various parts together. Nailing should be done correctly because even the highest grade of wood does not serve its purpose without proper nailing. Follow established rules in nailing the various wood members together. While most of the nailing will be described in future sections, Tables 1-2 and 1-3 list recommended practices used for framing and application of covering materials.

Most finish and siding nails have the same equivalent lengths. For example, an eightpenny common nail is the same length as an eightpenny galvanized siding nail, but not necessarily the same diameter.

LUMBER DESIGN VALUES

The design values that follow are for lumber of species and combinations of species manufactured and shipped by mills in the 12 western states. The do-it-yourselfer should have a general idea of what these design values are for use in all normal construction design.

Design values for visually graded lumber are assigned to six basic properties of wood. These are fiber stress in bending (fb), tension parallel to grain (ft), horizontal shear (Fv), compression parallel to grain (Fc), compression perpendicular to grain (Fci), and modules of elasticity (E).

Repetitive Member Design Values

In structures where 2-to-4-inch-thick lumber is used repetitively, for such things as joists, studs, rafters, and decking, the pieces side by side share the load and the strength of the entire assembly is enhanced. An example of the repetitive member design values for the fiber stress in bending "Fb" is shown in Table 1-4.

Table 1-2. Recommended Schedule for Nailing Framing and Sheathing.

Joining	Nailing method	Nails Number	Nails Size	Nails Placement
Header to joist	End-nail	3	16d	
Joist to sill or girder	Toenail	2-3	10d or 8d	
Header and stringer joist to sill	Toenail		10d	16 inches on center.
Bridging to joist	Toenail each end	2	8d	
Ledger strip to beam, 2 inches thick		3	16d	At each joist.
Subfloor, boards:				
1 by 6 inches and smaller		2	8d	To each joist.
1 by 8 inches		3	8d	To each joist.
Subfloor, plywood:				
At edges			8d	6 inches on center.
At intermediate joists			8d	8 inches on center.
Subfloor (2 by 6 inches, T&G) to joist or girder	Blind-nail (casing) and face-nail.	2	16d	
Soleplate to stud, horizontal assembly	End-nail	2	16d	At each stud.
Top plate to stud	End-nail	2	16d	
Stud to soleplate	Toenail	4	8d	
Soleplate to joist or blocking	Face-nail		16d	16 inches on center.
Doubled studs	Face-nail, stagger		10d	16 inches on center.
End stud of intersecting wall to exterior wall stud	Face-nail		16d	16 inches on center.
Upper top plate to lower top plate	Face-nail		16d	16 inches on center.
Upper top plate, laps and intersections	Face-nail	2	16d	
Continous header, 2 pieces, each edge			12d	12 inches on center.
Ceiling joist to top wall plates	Toenail	3	8d	
Ceiling joist laps at partition	Face-nail	4	16d	
Rafter to top plate	Toenail	2	8d	
Rafter to ceiling joist	Face-nail	5	10d	
Rafter to valley or hip rafter	Toenail	3	10d	
Ridge board to rafter	End-nail	3	10d	
Rafter to rafter through ridge board	Toenail	4	8d	
	Edge-nail	1	10d	
Collar beam to rafter:				
2-inch member	Face-nail	2	12d	
1-inch member	Face-nail	3	8d	
1-inch diagonal let-in brace to each stud and plate (4 nails at top).		2	8d	
Built-up corner studs:				
Studs to blocking	Face-nail	2	10d	Each side.
Intersecting stud to corner studs	Face-nail		16d	12 inches on center.
Built-up girders and beams, 3 or more members	Face-nail		20d	32 inches on center, each side.
Wall sheathing:				
1 by 8 inches or less, horizontal	Face-nail	2	8d	At each stud.
1 by 6 inches or greater, diagonal	Face-nail	3	8d	At each stud.
Wall sheathing, vertically applied plywood:				
⅜ inch and less thick	Face-nail		6d	6-inch edge.
½ inch and over thick	Face-nail		8d	12-inch intermediate.
Wall sheathing, vertically applied fiberboard:				
½ inch thick	Face-nail			1½-inch roofing nail.[1]
2⁵⁄₃₂ inch thick	Face-nail			1¾-inch roofing nail.[1]
Roof sheathing, boards, 4-, 6-, 8-inch width	Face-nail	2	8d	At each rafter.
Roof sheathing plywood:				
⅜ inch and less thick	Face-nail		6d	6-inch edge and 12-inch intermediate.
½ inch and over thick	Face-nail		8d	

[1] 3-inch edge and 6-inch intermediate.

Courtesy USDA

PRODUCT CLASSIFICATION

It is important to be specific when ordering materials. For the sake of clarity, identify product names of items such as paneling, structural decking, joists, rafters, studding, beams, and siding.

Be aware of all the species that can be used for the job. This broadens availability, which can lower costs. Check with your local supplier to find out which species are available.

Where wood color, grain, durability or other

Table 1-3. Nail sizes.

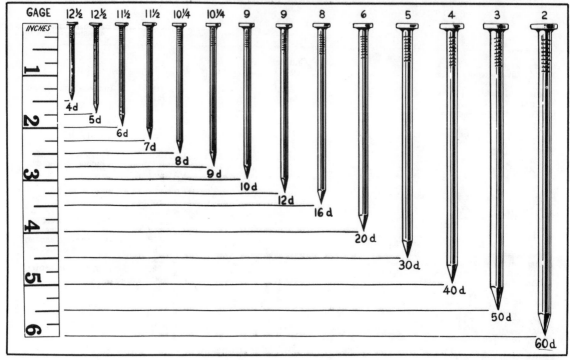

Courtesy USDA

special characteristics are important, select and specify the proper species accordingly.

Specify standard grades as described in Table 1-5. Consider all grades suitable for the intended use. When economy is a consideration, use the lowest grade suited to the specific job.

For standard products such as boards and framing, specify the nominal size by thickness and width in full inches. For example: 1 by 6 inches, 1 by 8 inches, 2 by 4 inches, 2 by 6 inches.

Indicate whether lumber is to have smooth surfaces, or if it is to be rough-or saw-textured.

List pattern number for standard materials, and provide details of special patterns. Where needed, identify whether you want tongue-and-groove (T&G), shiplap (S/L), or other patterns or workings.

Specify "DRY" lumber to assure long product life, increased nail-holding power, and improved paintability and workability. DRY refers to both major methods of drying lumber: air and kiln.

The relatively low thermal conductivity or *k,* of western softwoods provides for significant insulation

value. The k refers to the amount of heat (BTUs) transferred in 1 hour through 1 square foot of material 1 inch thick with a difference in temperature of 1 degree F.

The thermal conductivity of wood increases with a higher moisture content and with higher density.

Table 1-6 indicates the dimensions of all species and their usage. Table 1-7 gives standard lumber sizes, both nominal and dressed, based on WWPA rules.

MAKING STRONG WOOD JOINTS

Strong and inconspicuous joints in wood are necessary throughout construction. The following tips can help you do neater, more satisfactory, and more professional woodworking, and with far less waste.

Corner Joints

In woodworking, you will often need to make strong corner joints. There are various types of corner joints.

18

Table 1-4. Design Values/WWPA Standard Grading Rules.

LIGHT FRAMING—2″ Thick, 2″ Wide
Design Values in Pounds Per Square Inch*
Horizontal Shear "Fv", Compression Perpendicular "Fc ⊥" and Modulus of Elasticity "E" values are shown in Table 1, Light Framing.

Grades Described in Section 40.00.
Also Stress Rated Boards.
See Section 30.60 WWPA Grading Rules.

Species or Group	Grade	Extreme Fiber Stress in Bending "Fb"		Tension Parallel to Grain "Ft"	Compression Parallel to Grain "Fc"
		Single	Repetitive		
DOUGLAS FIR-LARCH	Construction	950	1100	500	1150
	Standard	450	500	225	925
	Utility	125	150	75	375
DOUGLAS FIR SOUTH	Construction	900	1050	475	1000
	Standard	425	475	225	850
	Utility	125	150	75	350
HEM-FIR	Construction	750	875	400	925
	Standard	350	400	175	775
	Utility	100	125	50	325
MOUNTAIN HEMLOCK	Construction	800	925	425	900
	Standard	375	425	200	725
	Utility	125	125	50	300
MOUNTAIN HEMLOCK-HEM-FIR	Construction	750	875	400	900
	Standard	350	400	175	725
	Utility	100	125	50	300
WESTERN HEMLOCK	Construction	825	950	450	1050
	Standard	375	450	200	850
	Utility	125	125	75	350
ENGELMANN SPRUCE—ALPINE FIR (Engelmann Spruce-Lodgepole Pine)	Construction	625	725	325	675
	Standard	300	325	150	550
	Utility	100	100	50	225
LODGEPOLE PINE	Construction	700	800	375	800
	Standard	325	375	175	675
	Utility	100	125	50	275
PONDEROSA PINE-SUGAR PINE (Ponderosa Pine Lodgepole Pine)	Construction	650	750	350	775
	Standard	300	350	150	625
	Utility	100	100	50	250
IDAHO WHITE PINE	Construction	600	700	325	775
	Standard	275	325	150	650
	Utility	75	100	50	275
WESTERN CEDARS	Construction	700	800	375	850
	Standard	325	375	175	700
	Utility	100	125	50	300
WHITE WOODS (Western Woods)	Construction	600	700	325	675
	Standard	275	325	150	550
	Utility	75	100	50	225

*These design values were calculated in accordance with ASTM standards. For information about use of these values, see Sections 100.00 through 170.00 in WWPA Grading Rules.

Courtesy Western Wood Products

The butt joint is the most commonly used. This involves simply nailing or screwing the end of one piece of wood to the end of the other. It is easy, fast, and effective. However, it also can't be used for many types of end joints because it leaves the head of the screws or nails exposed. If this is not a problem, this joint will do the job well. Also, the heads of the screws or nails can be countersunk and then covered with water putty or wood filler.

The dowel joint is basically the same as the butt joint except dowels rather than screws and nails are used to hold the two pieces of wood together. The dowel joint is made by drilling holes completely through one piece of wood and into the other. Dowels are then driven completely through one piece of wood and deeply into the other, and glued firmly into position to provide strength and to prevent slippage.

Blind dowel joints are made by drilling the holes only part way into each piece of wood. The dowels are then driven into these holes and glued into position. In this case, the dowels cannot be seen.

Dowel joints have the advantage of being inconspicuous, but simple butt joints, using screws or nails, tend to be stronger.

The end lap joint is made by sawing halfway through each of two pieces of wood, then knocking out or sawing away half of this area. The two pieces of wood are then put together with screws, nails, corrugated nails, etc. When completed, the end lap joint provides a great deal of strength, but the heads of the nails or screws are exposed.

Table 1-5. Grade Selector Charts.

APPEARANCE GRADES	SELECTS	B & BETTER (IWP—SUPREME)* C SELECT (IWP—CHOICE) D SELECT (IWP—QUALITY)
	FINISH	SUPERIOR PRIME E
	PANELING	CLEAR (ANY SELECT OR FINISH GRADE) NO. 2 COMMON SELECTED FOR KNOTTY PANELING NO. 3 COMMON SELECTED FOR KNOTTY PANELING
	SIDING (BEVEL, BUNGALOW)	SUPERIOR PRIME
	BOARDS SHEATHING & FORM LUMBER	NO. 1 COMMON (IWP—COLONIAL) NO. 2 COMMON (IWP—STERLING) NO. 3 COMMON (IWP—STANDARD) NO. 4 COMMON (IWP—UTILITY) NO. 5 COMMON (IWP—INDUSTRIAL) ALTERNATE BOARD GRADES SELECT MERCHANTABLE CONSTRUCTION STANDARD UTILITY ECONOMY

SPECIFICATION CHECK LIST

☐ Grades listed in order of quality.
☐ Include all species suited to project.
☐ Specify lowest grade that will satisfy job requirement.
☐ Specify surface texture desired.
☐ Specify moisture content suited to project.
☐ Specify ⓦ grade stamp. For finish and exposed pieces, specify stamp on back or ends.

Western Red Cedar

FINISH PANELING AND CEILING	CLEAR HEART A B
BEVEL SIDING	CLEAR — V.G. HEART A — BEVEL SIDING B — BEVEL SIDING C — BEVEL SIDING

Courtesy Western Wood Products

Table 1-6. Dimension/All Species.

LIGHT FRAMING 2″ to 4″ Thick 2″ to 4″ Wide	CONSTRUCTION STANDARD UTILITY	This category for use where high strength values are **NOT** required; such as studs, plates, sills, cripples, blocking, etc.
STUDS 2″ to 4″ Thick 2″ to 6″ Wide 10′ and Shorter	STUD	An optional all-purpose grade limited to 10 feet and shorter. Characteristics affecting strength and stiffness values are limited so that the "Stud" grade is suitable for all stud uses, including load bearing walls.
STRUCTURAL LIGHT FRAMING 2″ to 4″ Thick 2″ to 4″ Wide	SELECT STRUCTURAL NO. 1 NO. 2 NO. 3	These grades are designed to fit those engineering applications where higher bending strength ratios are needed in light framing sizes. Typical uses would be for trusses, concrete pier wall forms, etc.
STRUCTURAL JOISTS & PLANKS 2″ to 4″ Thick 5″ and Wider	SELECT STRUCTURAL NO. 1 NO. 2 NO. 3	These grades are designed especially to fit in engineering applications for lumber five inches and wider, such as joists, rafters and general framing uses.

Timbers 5″ and thicker

BEAMS & STRINGERS 5″ and thicker Width more than 2″ greater than thickness	SELECT STRUCTURAL NO. 1 NO. 2** NO. 3**

POSTS & TIMBERS 5″ x 5″ and larger Width not more than 2″ greater than thickness	SELECT STRUCTURAL NO. 1 NO. 2** NO. 3**

**Design values are not assigned.

Courtesy Western Wood Products

Table 1-7. Standard Lumber Sizes Nominal and Dressed. Based on WWPA Rules.

Product	Description	Nominal Size		Dressed Dimensions		
		Thickness In.	Width In.	Thickness In.	Width In.	Lengths Ft.
SELECTS AND COMMONS S-DRY	S1S, S2S, S4S, S1S1E, S1S2E	4/4 5/4 6/4 7/4 8/4 9/4 10/4 11/4 12/4 16/4	2 3 4 5 6 7 8 and wider	3/4 1 5/32 1 13/32 1 19/32 1 13/16 2 3/32 2 3/8 2 9/16 2 3/4 3 3/4	1 1/2 2 1/2 3 1/2 4 1/2 5 1/2 6 1/2 3/4 Off nominal	6 ft. and longer in multiples of 1′ except Douglas Fir and Larch Selects shall be 4′ and longer with 3% of 4′ and 5′ permitted.
FINISH AND BOARDS S-DRY	S1S, S2S, S4S, S1S1E, S1S2E Only these sizes apply to Alternate Board Grades.	3/8 1/2 5/8 3/4 1 1 1/4 1 1/2 1 3/4 2 2 1/2 3 3 1/2 4	2 3 4 5 6 7 8 and wider	5/16 7/16 9/16 5/8 3/4 1 1 1/4 1 3/8 1 1/2 2 2 1/2 3 3 1/2	1 1/2 2 1/2 3 1/2 4 1/2 5 1/2 6 1/2 3/4 off nominal	3′ and longer. In Superior grade, 3% of 3′ and 4′ and 7% of 5′ and 6′ are permitted. In Prime grade, 20% of 3′ to 6′ is permitted.
RUSTIC AND DROP SIDING	(D & M) If 3/8″ or 1/2″ T & G specified, same over-all widths apply. (Shiplapped, 3/8-in. or 1/2-in. lap)	1	6 8 10 12	23/32	5 3/8 7 1/8 9 1/8 11 1/8	4 ft. and longer in multiples of 1′
PANELING AND SIDING	T&G or Shiplap	1	6 8 10 12	23/32	5 7/16 7 1/8 9 1/8 11 1/8	4 ft. and longer in multiples of 1′
CEILING AND PARTITION	T&G	5/8 1	4 6	9/16 23/32	3 3/8 5 3/8	4 ft. and longer in multiples of 1′
BEVEL SIDING	Bevel or Bungalow Siding Western Red Cedar Bevel Siding available in 1/2″, 5/8″, 3/4″ nominal thickness. Corresponding thick edge is 15/32″, 9/16″ and 3/4″. Widths for 8″ and wider, 1/2″ off nominal.	1/2 3/4	4 5 6 8 10 12	15/32 butt, 3/16 tip 3/4 butt, 3/16 tip	3 1/2 4 1/2 5 1/2 7 1/4 9 1/4 11 1/4	3 ft. and longer in multiples of 1′ 3 ft. and longer in multiples of 1′
				Surfaced Dry / Green	Surfaced Dry / Green	
STRESS RATED BOARDS	S1S, S2S, S4S, S1S1E, S1S2E	1 1 1/4 1 1/2	2 3 4 5 6 7 8 and Wider	3/4 / 25/32 1 / 1 1/32 1 1/4 / 1 9/32	1 1/2 / 1 9/16 2 1/2 / 2 9/16 3 1/2 / 3 9/16 4 1/2 / 4 5/8 5 1/2 / 5 5/8 6 1/2 / 6 5/8 Off 3/4 / Off 1/2	6 ft. and longer in multiples of 1′

See coverage estimator chart below for dressed Shiplap and Tongue and Groove (T&G) widths.

MINIMUM ROUGH SIZES **Thicknesses and Widths Dry or Unseasoned All Lumber**
80% of the pieces in a shipment shall be at least 1/8″ thicker than the standard surfaced size, the remaining 20% at least 3/32″ thicker than the surfaced size. Widths shall be at least 1/8″ wider than standard surfaced widths.
When specified to be full sawn, lumber may not be manufactured to a size less than the size specified.

Courtesy Western Wood Products

The through mortise and tenon joint is easy to make with a power saw and a dado head. A slot is sawed into one piece of wood, while the end of the other piece of wood is notched out to fit the slot in the first piece. The notched piece of wood is then inserted into the slotted piece of wood where it can be glued, nailed, or screwed into position. When making a through mortise and tenon joint, be sure to measure the area to be notched and slotted before making the cuts.

The open mortise and tenon joint is made by cutting the slot or mortise only part way into one piece of wood. Then a notched-out area is made on the other piece that fits into the slotted area in the first piece. The open mortise and tenon cut creates a joint that is stronger than the through mortise and tenon joint. Though it can be easily cut with a mortising chisel on a drill press, it is a little more difficult to make than the through mortise and tenon joint. However, with a little experience and the

Table 1-8. Use of Nails by Size and Type.

Size	Lgth (in.)	Diam (in.)	Remarks	Where used
2d	1	.072	Small head	Finish work, shop work.
2d	1	.072	Large flathead	Small timber, wood shingles, lathes.
3d	1¼	.08	Small head	Finish work, shop work.
3d	1¼	.08	Large flathead	Small timber, wood shingles, lathes.
4d	1½	.098	Small head	Finish work, shop work.
4d	1½	.098	Large flathead	Small timber, lathes, shop work.
5d	1¾	.098	Small head	Finish work, shop work.
5d	1¾	.098	Large flathead	Small timber, lathes, shop work.
6d	2	.113	Small head	Finish work, casing, stops, etc., shop work.
6d	2	.113	Large flathead	Small timber, siding, sheathing, etc., shop work.
7d	2¼	.113	Small head	Casing, base, ceiling, stops, etc.
7d	2¼	.113	Large flathead	Sheathing, siding, subflooring, light framing.
8d	2½	.131	Small head	Casing, base, ceiling, wainscot, etc., shop work.
8d	2½	.131	Large flathead	Sheathing, siding, subflooring, light framing, shop work.
8d	1¼	.131	Extra-large flathead	Roll roofing, composition shingles.
9d	2¾	.131	Small head	Casing, base, ceiling, etc.
9d	2¾	.131	Large flathead	Sheathing, siding, subflooring, framing, shop work.
10d	3	.148	Small head	Casing, base, ceiling, etc., shop work.
10d	3	.148	Large flathead	Sheathing, siding, subflooring, framing, shop work.
12d	3¼	.148	Large flathead	Sheathing, subflooring, framing.
16d	3½	.162	Large flathead	Framing, bridges, etc.
20d	4	.192	Large flathead	Framing, bridges, etc.
30d	4½	.207	Large flathead	Heavy framing, bridges, etc.
40d	5	.225	Large flathead	Heavy framing, bridges, etc.
50d	5½	.244	Large flathead	Extra-heavy framing, bridges, etc.
60d	6	.262	Large flathead	Extra-heavy framing, bridges, etc.

[1] This chart applies to wire nails, although it may be used to determine the length of cut nails.

proper tools, you can make either of the joints easily.

The conventional miter joint is widely used for making corners in various types of woodwork. It is not recommended where the joint is to be subject to excessive weight or strain. The conventional miter joint is made by mitering each corner at a 45-degree angle. You will need a regular miter box or a homemade miter box if it becomes necessary to cut many joints of this type. Nails, screws, or corrugated nails can be used for attaching the two pieces of wood when making a conventional miter joint. The conventional miter joint is often used for making trims around cabinet doors and on other trim pieces.

A miter joint with a pline adds a great deal of strength to a common miter joint. This can easily be made with ordinary tools. It requires simply cutting a regular 45-degree angle miter joint and then cutting a groove in each end of the pieces to be mitered. After sawing the grooves, a spline is sawed to fit the grooves that have been created. The spline in the mitered joint can be held in position with any top grade adhesive. It can also be nailed or screwed into position if appearance is not important.

Joining Top and Side Pieces

In working with wood, all joints will not be corner joints. You will occasionally need to joint a top piece of wood to a side piece. Again, the standard butt joint is the most commonly used joint. It can be nailed or screwed together if appearance is not important, and provides a strong joint that is completely satisfactory for ordinary jobs.

If you are an experienced handyman, you might want to use the lock miter joint for joining a top or bottom to side pieces of wood. You will need a power saw to make the lock miter joint. Accuracy is important when sawing the lock miter joint. When correctly sawed and grooved together, the lock miter joint provides a strong and inconspicuous joint.

The mitered rabbet joint is similar to the lock miter joint, and it too must be made with power equipment. Accuracy in sawing and rabbeting is very important. The two pieces of wood on any rabbet joint can be held together with screws, nails, adhesives or dowels. How the two pieces of wood are held together will depend largely on the strength required at the joint and the objections to nail or screw

22

heads. Regardless of how the mitered rabbet joint is secured, it provides an excellent joint with a professional look and a great deal of strength.

The regular rabbet joint is much easier to make than the mitered rabbet joint. Although power equipment is helpful, the regular rabbet joint can be made with ordinary hand tools. The rabbet can be cut either into the side piece or the top piece when two pieces of wood are joined.

Where you position the rabbet cut depends largely on where you want the half section of grained end to appear. With a rabbet joint, the grained end of one piece of wood is completely hidden.

The box corner joint is one that should be undertaken only by an experienced do-it-yourselfer or professional carpenter. It requires sawing a groove in one piece of wood and a tongue or flange in the other. Such a joint probably won't be required because the variety of others available will suffice most types of jointing. If you wish to try, however, it does provide a strong joint that can be held together with adhesive, nails, or screws. It leaves no end grain showing.

The milled corner joint also creates a corner with no end grain visible, which is highly desirable on some types of woodwork. The mitered corner joint is often used when making drawers. It creates a joint that is much stronger than the box corner joint and less subject to cracking. Milled corner joints require accurately sawing the tongues and grooves. You will need power equipment for making a joint of this type.

The half blind dovetail joint is used almost exclusively for making drawers. It is a joint that requires some experience and good power tools. The half blind dovetail joint can be held together with adhesives and is an excellent joint with no end grain visible.

A complete open dovetail joint can be made by simply cutting through the second piece of wood. This provides a joint that is equally strong with the half blind dovetail joint, but the end grain is visible on both sides of the joint.

Joining One Board in the Center of Another

Some woodworking jobs will require a joint to be

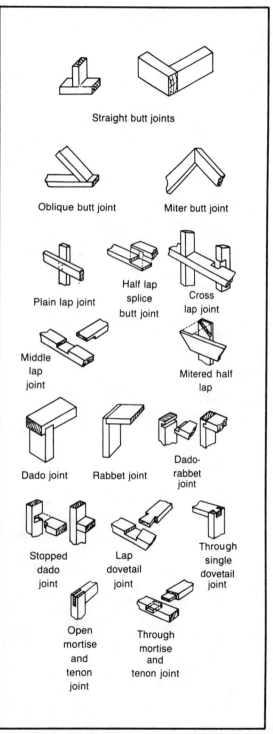

Fig. 1-17. Various Wood cuts. Courtesy USDA

23

made where the end of one board butts against the center of another. Again, you have the choice of several joints.

The regular butt joint is again the most commonly used. It can be secured by using nails, screws or adhesives. A butt joint provides a lot of strength, and if the visible heads of nails or screws are not objectionable it will do the job well.

The dado joint requires a slot to be cut into one piece of wood to receive the end of the other. The dado joint is much stronger than the butt joint and also creates a more professional appearance. A dado can be cut into the board with a dado head on a power saw, a regular hand saw, a dado plane, or even a chisel and a mallet. Wood screws, nails, or dowels can be used to hold dado joints of this type.

The stopped dado joint is a modified version of the regular dado joint. The stopped dado joint is a little more difficult to make by hand, but is quite easily done if you have the proper power tools. It creates a neater appearance than the regular dado joint because the front edge is uncut. Thus, the slotted area is not visible from the front side. Ordinarily wood adhesives, wood screws, nails, or dowels can be used to hold the two pieces of wood together to make a stopped dado joint.

Figure 1-17 illustrates the joints discussed above.

NAILS

Nails are the fasteners used for frame construction. There are many types of nails, classified according to use and form. The wire nail is round, straight, pointed, and varies in size, weight, shape of head, type of point, and finish. Follow these rules when using nails:

The nail should be at least three times as long as the thickness of wood it is intended to hold. Two-thirds of the length of the nail is driven into the second piece for proper anchorage; one-third anchors the piece being fastened.

Nails should be driven at an angle slightly toward each other and placed to provide the greatest holding power. Nails driven with the grain do not hold as well as nails driven across the grain.

Nails are the cheapest and easiest fasteners to

be used. Screws of comparable size provide more holding power; bolts provide still more.

Common Wire Nails

Common wire nails and box nails are the same except that the wire sizes are one or two numbers smaller for a given length of the box nail than they are for the common nail. The common wire nail is used for house framing.

Finishing Nails

The finishing nail is made from finer wire and has a smaller head than the common nail. It may be set below the surface of the wood, which leaves only a small hole that can be easily filled with putty. It is generally used for interior or exterior finishing work and for finished carpentry and cabinetmaking.

Scaffold or Form Nails

The scaffold, form, or staging nail appears to have two heads. The lower head (shoulder) permits the nail to be driven securely home while the upper head projects above the wood to make it easy to pull. The scaffold nail is not meant to be permanent.

Roofing Nails

Roofing nails are round-shafted, diamond-pointed, galvanized nails that are relatively short length with large heads. They fasten flexible roofing and resist continuous exposure to weather. If shingles or roll roofing is put on over old roofing, the roofing nails

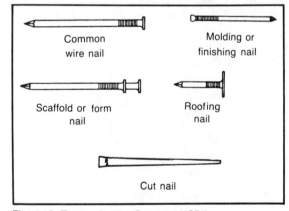

Fig. 1-18. Types of nails. Courtesy USDA

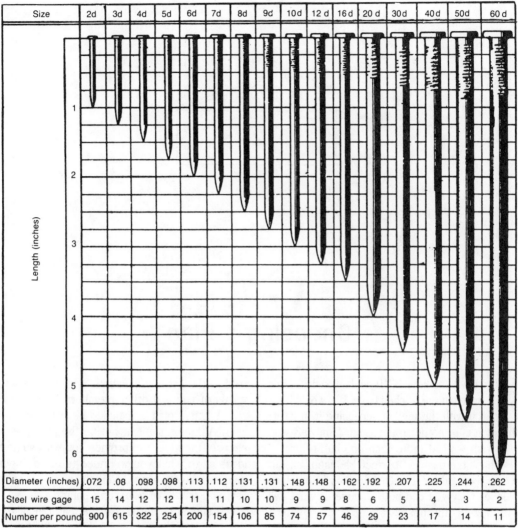

Size	2d	3d	4d	5d	6d	7d	8d	9d	10d	12 d	16 d	20 d	30 d	40 d	50d	60 d
Diameter (inches)	.072	.08	.098	.098	.113	.112	.131	.131	.148	.148	.162	.192	.207	.225	.244	.262
Steel wire gage	15	14	12	12	11	11	10	10	9	9	8	6	5	4	3	2
Number per pound	900	615	322	254	200	154	106	85	74	57	46	29	23	17	14	11

Fig. 1-19. Nail dimensions and details. Courtesy USDA

must be long enough to go through the old materials and secure the new. Asphalt roofing must be fastened with corrosion-resistant nails, never with plain nails. Nailing is begun in the center of the shingle, just above the cutouts or slots to avoid buckling.

Cut Nails

Cut nails are wedge-shaped with a head on the large end. They are often used to nail flooring because they are of very hard steel and have good holding power.

Figure 1-18 illustrates the various nails.

Nail sizes are designated by the term "penny". This term applies to the length of the nail (1 penny, 2 penny, etc.). The approximate number of nails per pound varies according to the size and type. The wire gauge number varies according to type. Fig. 1-19, the "d" next to the numbers in the "size" column is the abbreviation of "penny" and should be read "2 penny", "3 penny", etc. Table 1-8 provides a reference for the sizes and types of nails, and the uses they serve.

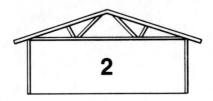

2

Choosing a Plan

How often have you heard: ''If we had to do it over we would have put the kitchen here. We would have built a porch with the idea of screening it in later, or turning it into another room as our family grew.'' Or: ''I wish we had more than one bathroom!''

Perhaps there is no ''perfect'' house plan, but there are hundreds of standard designs from which to choose. Only you can be the judge of what will suit your needs the best.

Are you wondering where to begin? First, make a written list of the following questions, provide your own answers, and you are on your way:

—Are you retired?
—Newlyweds?
—How many children do you have?
—How many children do you plan to have?
—Do you need a house that can be ''expanded?''
—Are you indoor/outdoor people?
—Are there any incapacities healthwise?
—How much can you afford?
—Other personal questions?

Add to the list any questions which come into the minds of you and other members of the family.

A house plan is included in this chapter (Fig. 2-1). Although it might not be the plan you would choose, the idea is to provide you with some basic ideas as to what you will need to decide upon for your own house. By studying the plans and materials list, you will have a starting point.

It should be noted that the materials listed are the minimum amount required for the particular job.

THE HOUSE PLAN

The house plan in this chapter is an expandable type. With its steeply pitched roof, there is more than adequate space on the second floor for two dormitory-type bedrooms, which can accommodate up to eight children. The working drawings also provide for an additional bath on the second floor if required. The house is 24 by 32 feet in size with an area of 768 square feet on the first floor and about 460 square feet of usable space on the second. The first floor contains a moderate-size living room, a compact

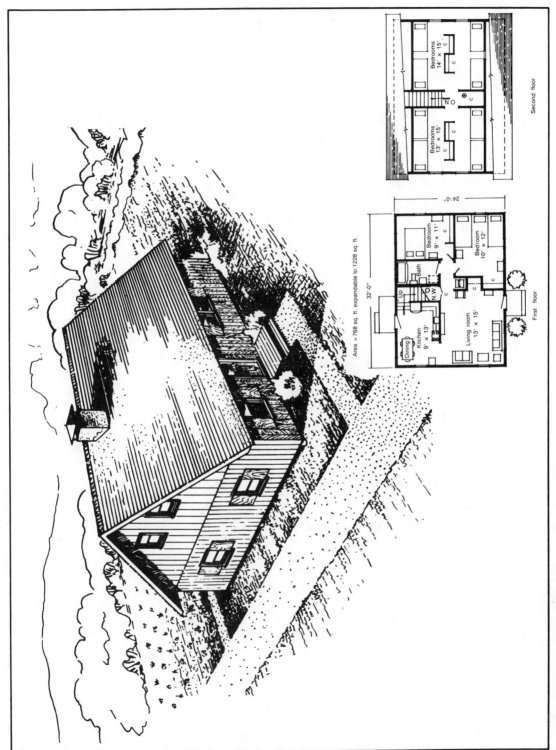

Area = 768 sq. ft. expandable to 1228 sq. ft.

Second floor

Bedrooms
14' x 15'

Bedrooms
13' x 15'

First floor

24'-0"

32'-0"

Bedroom
9' x 11'

Bedroom
10' x 12'

Bath

Up

Kitchen
9' x 13'

Dining

Living room
13' x 15'

Fig. 2-1. Diagram of House Plan Design #1. Courtesy USDA

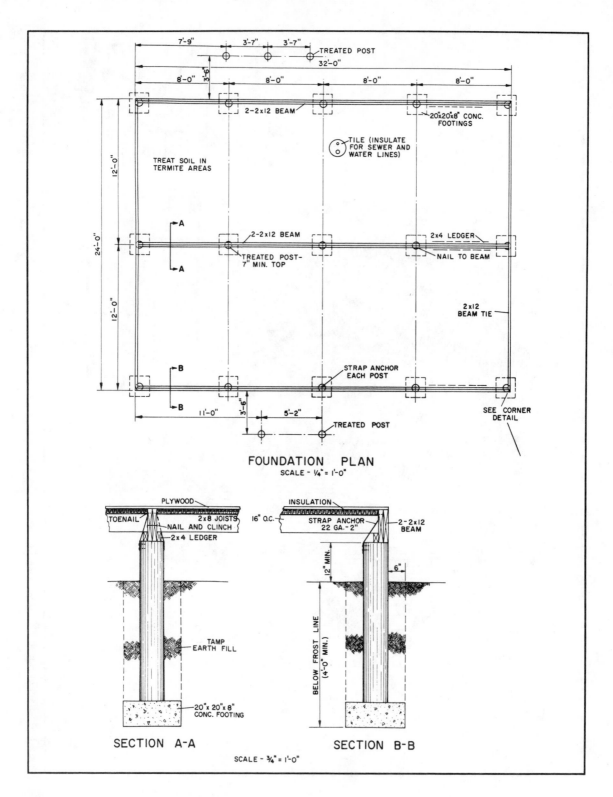

FOUNDATION PLAN

SCALE - ¼" = 1'-0"

7'-9"   3'-7"   3'-7"   TREATED POST

32'-0"

8'-0"   8'-0"   8'-0"   8'-0"

3'-6"

2-2x12 BEAM

20"x20"x8" CONC. FOOTINGS

TILE (INSULATE FOR SEWER AND WATER LINES)

TREAT SOIL IN TERMITE AREAS

12'-0"

24'-0"

A

A

2-2x12 BEAM

2x4 LEDGER

NAIL TO BEAM

TREATED POST- 7" MIN. TOP

12'-0"

2x12 BEAM TIE

B

B

STRAP ANCHOR EACH POST

SEE CORNER DETAIL

11'-0"   3'-6"   5'-2"

TREATED POST

SECTION A-A

PLYWOOD

TOENAIL

2x8 JOISTS

NAIL AND CLINCH

2x4 LEDGER

TAMP EARTH FILL

20"x 20"x 8" CONC. FOOTING

SECTION B-B

INSULATION

16" O.C.

STRAP ANCHOR 22 GA.- 2"

2-2x12 BEAM

12" MIN.

6"

BELOW FROST LINE (4'-0" MIN.)

SCALE - ¾" = 1'-0"

28

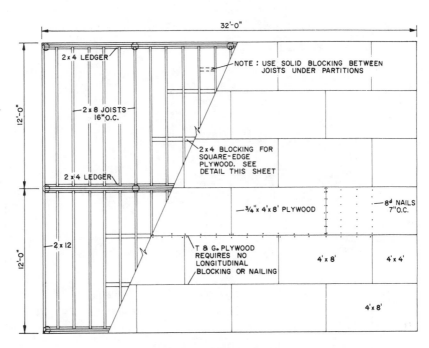

32'-0"

2 x 4 LEDGER

NOTE : USE SOLID BLOCKING BETWEEN JOISTS UNDER PARTITIONS

2 x 8 JOISTS 16" O.C.

2 x 4 BLOCKING FOR SQUARE-EDGE PLYWOOD. SEE DETAIL THIS SHEET

2 x 4 LEDGER

8ᵈ NAILS 7" O.C.

¾" x 4' x 8' PLYWOOD

12'-0"

2 x 12

T & G. PLYWOOD REQUIRES NO LONGITUDINAL BLOCKING OR NAILING

4' x 8'

4' x 4'

12'-0"

4' x 8'

FLOOR FRAMING PLAN
SCALE - ¼" = 1'-0"

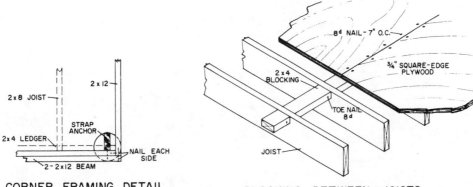

8ᵈ NAIL - 7" O.C.

2 x 4 BLOCKING

¾" SQUARE-EDGE PLYWOOD

TOE NAIL 8ᵈ

JOIST

2 x 12

2 x 8 JOIST

STRAP ANCHOR

2 x 4 LEDGER

NAIL EACH SIDE

2 - 2 x 12 BEAM

CORNER FRAMING DETAIL
SCALE - 1" = 1'-0"

BLOCKING BETWEEN JOISTS FOR SQUARE-EDGE PLYWOOD
(NO SCALE)

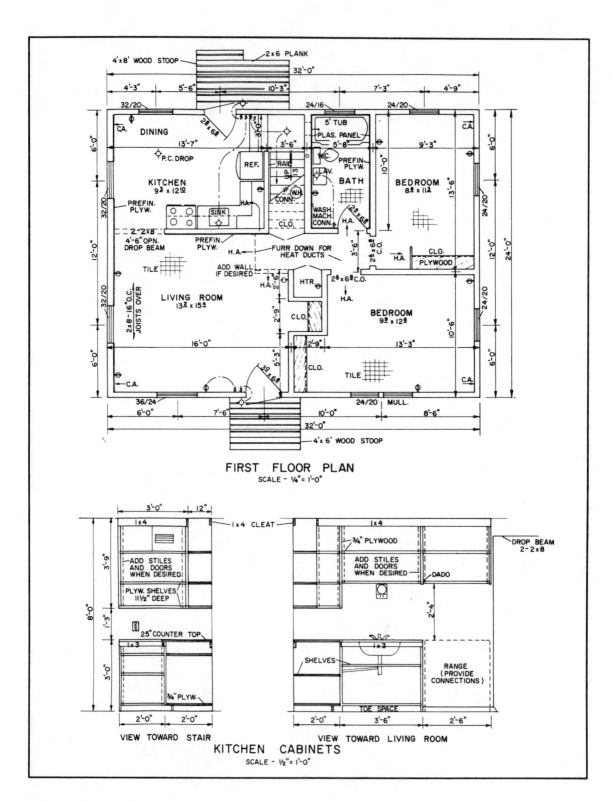

FIRST FLOOR PLAN
SCALE - ¼" = 1'-0"

KITCHEN CABINETS
SCALE - ½" = 1'-0"

VIEW TOWARD STAIR

VIEW TOWARD LIVING ROOM

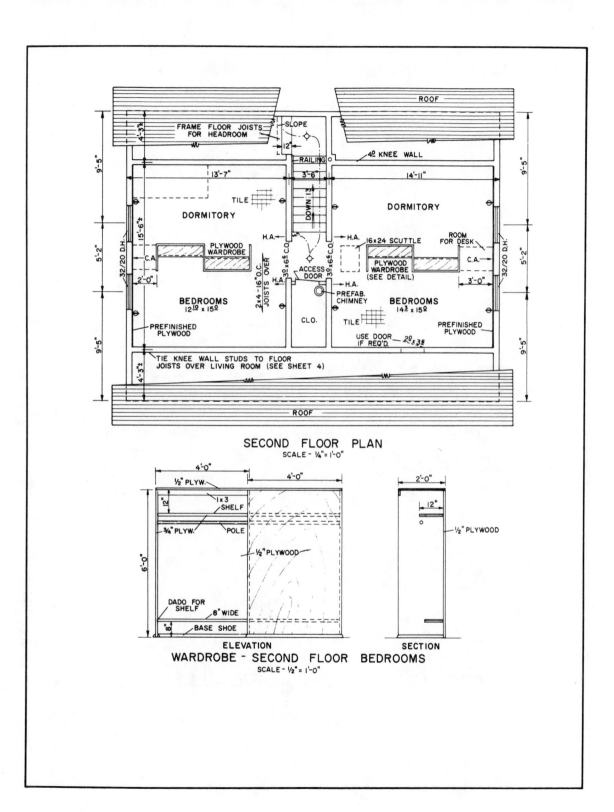

SECOND FLOOR PLAN
SCALE - ¼" = 1'-0"

ELEVATION
WARDROBE - SECOND FLOOR BEDROOMS
SCALE - ½" = 1'-0"

SECTION

31

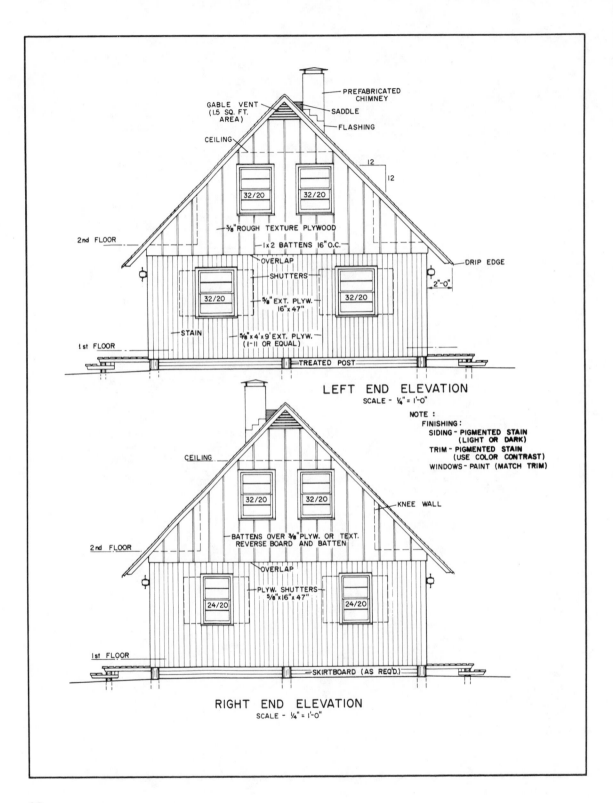

PREFABRICATED
CHIMNEY

GABLE VENT
(1.5 SQ. FT.
AREA)

SADDLE

FLASHING

CEILING

12
12

2nd FLOOR

32/20   32/20

⅜" ROUGH TEXTURE PLYWOOD

1x2 BATTENS 16" O.C.

OVERLAP

SHUTTERS

DRIP EDGE

2"-0"

32/20   ⅝" EXT. PLYW.   32/20
            16" x 47"

STAIN

1st FLOOR

⅝" x 4' x 9' EXT. PLYW.
(I-II OR EQUAL)

TREATED POST

LEFT END ELEVATION
SCALE - ¼" = 1'-0"

NOTE :
FINISHING :
SIDING - PIGMENTED STAIN
(LIGHT OR DARK)
TRIM - PIGMENTED STAIN
(USE COLOR CONTRAST)
WINDOWS - PAINT (MATCH TRIM)

CEILING

32/20   32/20

KNEE WALL

BATTENS OVER ⅜" PLYW. OR TEXT.
REVERSE BOARD AND BATTEN

2nd FLOOR

OVERLAP

24/20   PLYW. SHUTTERS   24/20
          ⅝" x 16" x 47"

1st FLOOR

SKIRTBOARD (AS REQ'D.)

RIGHT END ELEVATION
SCALE - ¼" = 1'-0"

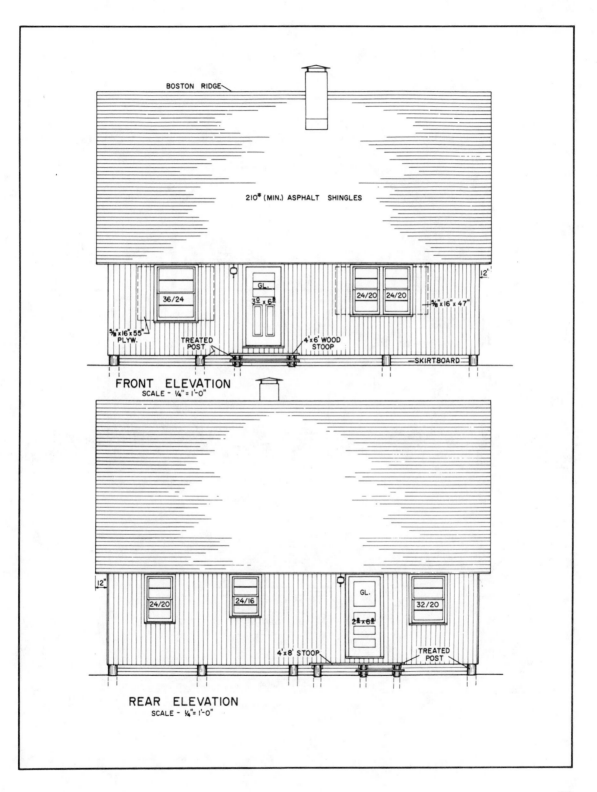

BOSTON RIDGE

210# (MIN.) ASPHALT SHINGLES

12'

36/24

GL.
3⁰ x 6⁸

24/20   24/20

⅝"x16"x47"

⅝"x16"x55"
PLYW.

TREATED
POST

4'x6' WOOD
STOOP

SKIRTBOARD

FRONT ELEVATION
SCALE - ¼"= 1'-0"

12'

24/20   24/16

GL.
2⁸x6⁸

32/20

4'x8' STOOP

TREATED
POST

REAR ELEVATION
SCALE - ¼"= 1'-0"

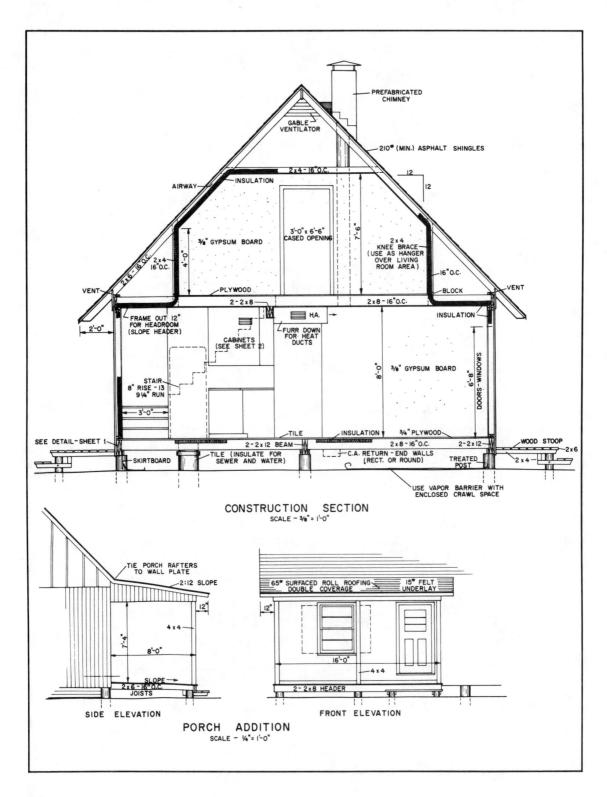

PREFABRICATED
CHIMNEY

GABLE
VENTILATOR

210# (MIN.) ASPHALT SHINGLES

2 x 4 - 16" O.C.

AIRWAY — INSULATION

12

12

⅜" GYPSUM BOARD

3'-0" x 6'-6"
CASED OPENING

7'-6"

2 x 4
KNEE BRACE
(USE AS HANGER
OVER LIVING
ROOM AREA)

2 x 4
16" O.C.

2 x 6 - 16" O.C.

4'-0"

16" O.C.

VENT

PLYWOOD

2 - 2 x 8

2 x 8 - 16" O.C.

BLOCK

VENT

FRAME OUT 12"
FOR HEADROOM
(SLOPE HEADER)

2'-0"

H.A.

FURR DOWN
FOR HEAT
DUCTS

INSULATION

CABINETS
(SEE SHEET 2)

STAIR
8" RISE - 13
9¼" RUN

8'-0"

⅜" GYPSUM BOARD

6'-8"

DOORS - WINDOWS

3'-0"

TILE

INSULATION

¾" PLYWOOD

SEE DETAIL - SHEET 1

2 - 2 x 12 BEAM

SKIRTBOARD

TILE (INSULATE FOR
SEWER AND WATER)

C.A. RETURN - END WALLS
(RECT. OR ROUND)

2 x 8 - 16" O.C.

TREATED
POST

2 - 2 x 12

WOOD STOOP

2 x 6

2 x 4

USE VAPOR BARRIER WITH
ENCLOSED CRAWL SPACE

CONSTRUCTION SECTION
SCALE – ⅜" = 1'-0"

TIE PORCH RAFTERS
TO WALL PLATE

2:12 SLOPE

12"

4 x 4

7'-4"

8'-0"

SLOPE

2 x 6 - 16" O.C.
JOISTS

SIDE ELEVATION

65# SURFACED ROLL ROOFING
DOUBLE COVERAGE

15# FELT
UNDERLAY

12"

16'-0"

4 x 4

2 - 2 x 8 HEADER

FRONT ELEVATION

PORCH ADDITION
SCALE – ¼" = 1'-0"

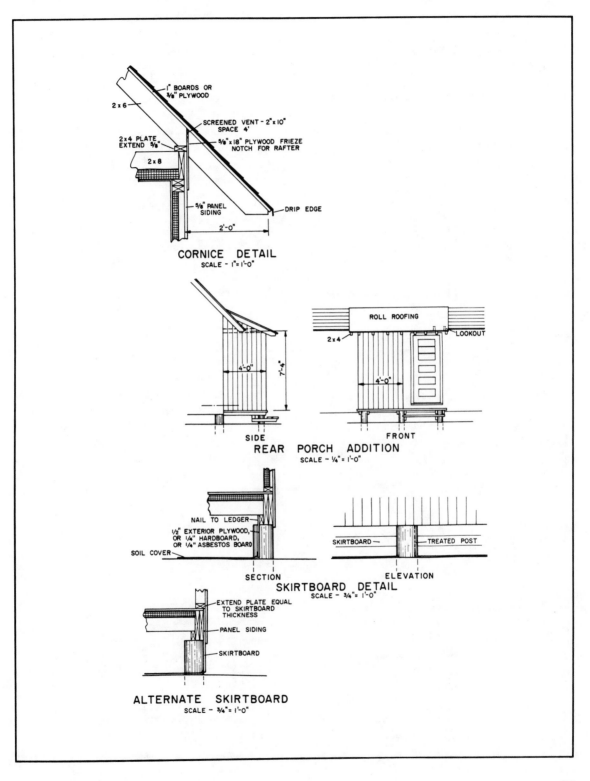

CORNICE DETAIL
SCALE – 1"= 1'-0"

1" BOARDS OR 3/8" PLYWOOD

2 x 6

SCREENED VENT – 2"x 10"
SPACE 4'

2 x 4 PLATE
EXTEND 5/8"

5/8"x 18" PLYWOOD FRIEZE
NOTCH FOR RAFTER

2 x 8

5/8" PANEL
SIDING

DRIP EDGE

2'-0"

REAR PORCH ADDITION
SCALE – 1/4"= 1'-0"

SIDE

4'-0"

7'-4"

FRONT

ROLL ROOFING

2 x 4

LOOKOUT

4'-0"

SKIRTBOARD DETAIL
SCALE – 3/4"= 1'-0"

SECTION

NAIL TO LEDGER

1/2" EXTERIOR PLYWOOD,
OR 1/4" HARDBOARD,
OR 1/4" ASBESTOS BOARD

SOIL COVER

ELEVATION

SKIRTBOARD

TREATED POST

ALTERNATE SKIRTBOARD
SCALE – 3/4"= 1'-0"

EXTEND PLATE EQUAL
TO SKIRTBOARD
THICKNESS

PANEL SIDING

SKIRTBOARD

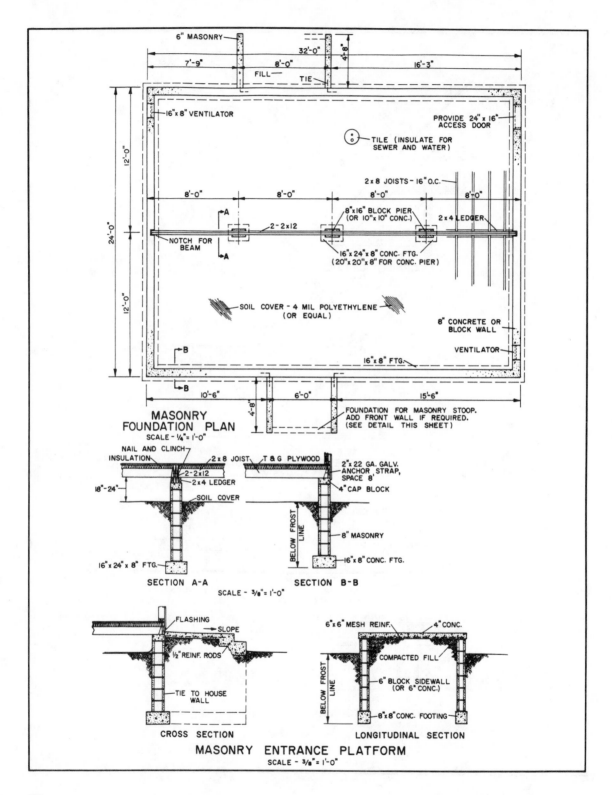

6" MASONRY

32'-0"

4'-8"

7'-9"   8'-0"   16'-3"

FILL

TIE

16" x 8" VENTILATOR

PROVIDE 24" x 16"
ACCESS DOOR

TILE (INSULATE FOR
SEWER AND WATER)

12'-0"

2 x 8 JOISTS - 16" O.C.

8'-0"   8'-0"   8'-0"   8'-0"

24'-0"

8"x16" BLOCK PIER
(OR 10"x10" CONC.)

2 x 4 LEDGER

A   A

2 - 2 x 12

NOTCH FOR
BEAM

16"x 24" x 8" CONC. FTG.
(20"x 20"x 8" FOR CONC. PIER)

12'-0"

SOIL COVER - 4 MIL POLYETHYLENE
(OR EQUAL)

8" CONCRETE OR
BLOCK WALL

VENTILATOR

B   B

16"x 8" FTG.

10'-6"   6'-0"   15'-6"

4'-8"

MASONRY
FOUNDATION PLAN
SCALE - 1/4" = 1'-0"

FOUNDATION FOR MASONRY STOOP.
ADD FRONT WALL IF REQUIRED.
(SEE DETAIL THIS SHEET)

NAIL AND CLINCH
INSULATION

2 x 8 JOIST   T & G PLYWOOD

2 - 2 x 12

2 x 4 LEDGER

18"-24"

SOIL COVER

16" x 24" x 8" FTG.

SECTION A-A

2 x 22 GA. GALV.
ANCHOR STRAP,
SPACE 8'

4" CAP BLOCK

BELOW FROST LINE

8" MASONRY

16"x 8" CONC. FTG.

SECTION B-B

SCALE - 3/8" = 1'-0"

FLASHING

SLOPE

1/2" REINF. RODS

TIE TO HOUSE
WALL

CROSS SECTION

6"x 6" MESH REINF.   4" CONC.

COMPACTED FILL

BELOW FROST LINE

6" BLOCK SIDEWALL
(OR 6" CONC.)

8"x 8" CONC. FOOTING

LONGITUDINAL SECTION

MASONRY ENTRANCE PLATFORM
SCALE - 3/8" = 1'-0"

36

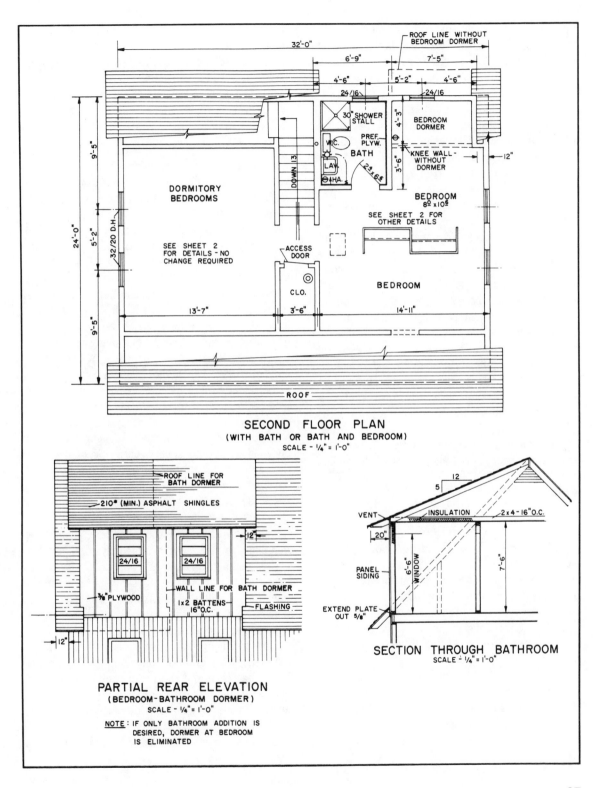

ROOF LINE WITHOUT
BEDROOM DORMER

32'-0"

6'-9"   7'-5"

4'-6"   5'-2"   4'-6"

24/16   24/16

30" SHOWER
STALL

BEDROOM
DORMER

9'-5"

PREF.
PLYW.

W.C.

BATH

KNEE WALL
WITHOUT
DORMER

12"

LAV.

DORMITORY
BEDROOMS

2'-6⁵

HA.

4'-3"

3'-6"

BEDROOM
8⁰x10⁵

DOWN 13

24'-0"

5'-2"

32/20 D.H.

SEE SHEET 2
FOR DETAILS - NO
CHANGE REQUIRED

SEE SHEET 2 FOR
OTHER DETAILS

ACCESS
DOOR

BEDROOM

9'-5"

CLO.

13'-7"   3'-6"   14'-11"

ROOF

SECOND FLOOR PLAN
(WITH BATH OR BATH AND BEDROOM)
SCALE - ¼" = 1'-0"

ROOF LINE FOR
BATH DORMER

210# (MIN.) ASPHALT SHINGLES

12"

5   12

24/16   24/16

VENT

INSULATION   2x4-16"O.C.

20"

⅜" PLYWOOD

WALL LINE FOR BATH DORMER

1x2 BATTENS
16"O.C.

FLASHING

PANEL
SIDING

WINDOW

6'-6"

7'-6"

12"

EXTEND PLATE
OUT 5/8"

PARTIAL REAR ELEVATION
(BEDROOM-BATHROOM DORMER)
SCALE - ¼" = 1'-0"

NOTE: IF ONLY BATHROOM ADDITION IS
DESIRED, DORMER AT BEDROOM
IS ELIMINATED

SECTION THROUGH BATHROOM
SCALE - ¼" = 1'-0"

37

kitchen with a large adjoining dining area, two bedrooms, and a bath. Storage space is adequate with four closets on the first floor and five on the second. The first-floor bath is arranged to accommodate a washing machine.

There are several important factors that aid in reducing the cost of this home. One is the fact that it is a crawlspace house, which eliminates the need for extensive excavation and grading. In addition, the floor framing is supported by wood foundation posts—treated for long life—resting on concrete footings. A more costly masonry foundation is, however, an alternative.

The use of a single covering material for the subfloor and the exterior walls also leads to reduced costs. The subfloor consists of tongue-and-groove plywood or square-edge plywood with edge blocking and serves as a base for a resilient floor covering.

The panel siding, with perimeter nailing, eliminates the need for corner bracing as well as the need for sheathing. Such coverings are usually rough-textured exterior-grade plywood, which can be finished with a pigmented stain. Suitable stains are available in many colors, and contrasts can be obtained by treating the trim and shutters with a different color.

An adequate forced-air heating unit with relatively short heat runs is also a part of the design. Insulation is included in the floor, wall, and ceiling areas.

Preliminaries

Excavation and Grading: Sod, growing plants, shrubs, stumps, and trees must be removed and the ground smoothed in building area and 2 feet outside of building line. Excavate footings to required depth and size required for the particular plan. No forms are required if soil is stable. Locate excavated soil conveniently for backfilling around treated wood posts.

Concrete Work: Footings are poured over undisturbed soil in excavations to the thickness and size required on plan. Top surface to be level.

If premixed concrete is available (and highly recommended for the do-it-yourselfer, unless you order yours premixed from your local supplier), use 5-bag mix.

Treated Wood Posts: Pressure-treated wood posts with 7-inch minimum top diameter must be treated to conform to Federal Specification TT-W-571.

Carpentry

This comprises all rough and finish carpentry necessary to complete the house shown in the plan, including layout, cutting and fitting framing, and other carpentry items.

Wood Framing: Dimension material for studs in this plan is to be standard, or third grade, material. Second grade is used for floor joists and framing, ceiling joists, rafters, beams, and truss construction. Douglas fir, southern pine, or an equivalent wood is specified. All floor and ceiling joists, studs, and rafters shall be spaced 16 inches on center. Moisture content of framing lumber is not to exceed 19 percent.

Subfloor: The subfloor is plywood to serve as a floor alone or as a base for resilient tile. The plywood is ¾-inch C-C plugged Exterior grade, touchsanded, and with matched edges in Douglas fir, southern pine, or an equivalent. When matched-edge plywood is not obtainable, square-edged plywood in the same grade and thickness can be substituted, but 2-by-2-inch blocking is used for all longitudinal joints. Toenail 2-by-2-inch blocks flatwise in each joist space.

Roof Sheathing: Roof sheathing shall be ⅜-inch Douglas fir or southern pine plywood in standard sheathing grade or nominal 1-by-6- or 1-by-4-inch boards in No. 3 Douglas fir, southern pine, or equivalent. Boards shall be square edge, shiplap, or dressed and matched, and laid up tight, at a moisture content of no more than 14 percent.

Insulation: Floors and walls of the first floor, walls and ceiling of the second floor, and the second floor area outside of the knee walls is to be insulated with standard batt- or blanket-type flexible insulation, with a vapor barrier placed toward the inside of the building. Unless otherwise specified, minimum insulation thicknesses for the central- and northern-tier states are: 3 inches in the ceiling, 2 inches in the wall and floor. The vapor barrier is to have a maximum perm value of 0.30.

Siding: Panel siding at gable ends shall be ⅜-inch thick. Remaining walls have panel siding ⅝-inch thick. Edges of the sheets should be dipped or brush-coated with a water-repellent preservative before installation. A pigmented stain finish is recommended. Nail plywood at each stud and at ends of panels with galvanized or other rust-resistant nails spaced 7 to 8 inches apart. Use sixpenny nails for ¾-inch plywood and eightpenny nails for the ⅝-inch plywood.

Exterior Millwork: Exterior trim and similar materials are of No. 2 Ponderosa pine or equivalent, suitable for staining.

Complete double-hung windows are used with the upper and lower sashes cut to two horizontal lights, glazed with single-strength glass. Units are treated with water-repellent preservative. Sash is to be furnished fully balanced and fitted with outside casing in place.

Exterior door frames have 1⅜-inch rabbeted jambs and 1⅝-inch oak or softwood sills with metal edges, all assembled.

Exterior doors are standard 1¾ inch, with solid stiles and rails. They are a panel type with glazed openings. Galvanized fly screen is used for gable end outlet ventilators and for inlet ventilators located in the plywood frieze board.

Interior door frames and cased openings are nominal 1-inch Ponderosa pine or equal in "D" select. Install stops only where doors are specified. Interior doors are 1⅜-inch thick, five-cross-panel style with solid stiles and rails.

Interior trim shall be Ponderosa pine or equal in "D" select in the following sizes:

Casings: 11/16″ by 2¼″ Stops: 7/16″ by 1⅜″ or wider
Base: 7/16″ by 2¼″
Base shoe: ½″ by ¾″ (when required)

All ceilings are to be ⅜-inch gypsum board with recessed edges, and with the length applied across the ceiling joists. End joints are staggered at least 16 inches. Walls are ⅜-inch gypsum board with recessed edges and applied vertically.

Walls of tub recess are covered with plastic-finished hardboard panels over gypsum board. Install with mastic in accordance with manufacturer's directions. Inside covers, edges, and tub edges are finished with plastic moldings.

Finish flooring throughout should be ⅛-inch-thick asphalt tile in 9-by-9-inch size, "B" quality. Combination plywood subfloor should be cleaned, nails driven flush, and joints sanded smooth where required. Apply tile in accordance with manufacturer's recommendations. Rubber baseboard is installed in the bathroom, wood base in the remainder of the house.

Furnish and install all rough and finish hardware as needed for perfect operation, i.e., locks, hinges, latches, cabinet pulls.

Sheet Metal Work and Roofing

Sheet metal flashing, when required for the prefabricated chimney and vent stack, should be 29-gauge galvanized iron or painted terneplate.

Roofing should be a minimum of 210-pound square-tab 12-by-36-inch asphalt shingles. A minimum of four ⅞-inch galvanized roofing nails shall be used for each 12-by-36-inch single strip. Correct any defects or leaks.

Electrical Work

The work includes all materials and labor necessary to make the systems complete as shown on the plan. All work and materials must comply with local requirements or those of the National Electrical Code. All wiring is concealed and carried in BX or other approved conduit to each outlet, switch, fixture, and appliance. Panel is 100-amp capacity with an overload cutout. Wall fixtures to consist of: Two outside wall fixtures with crystal glass, two overhead fixtures for kitchen and bath.

Heating

Heater and prefabricated chimney is installed with supply ducts located in the furred-down ceiling of hall. Heater shall be for LP or natural gas with a 100,000 minimum BTU input, or as required for each specific area. There is a cold air return at furnace base and from each corner of the house.

Plumbing

All plumbing must be installed in accordance with local or national plumbing codes. Hot-and cold-water connections are to be furnished to all fixtures as required. Sewer and water and gas lines (when required) extend to the building line with a water shutoff valve. When required, a framed and insulated box is used to protect water and sewer lines from freezing in the crawlspace.

The following fixtures are required in this plan: A kitchen sink, bathtub, water closet, lavatory, hot water heater, and washing machine.

Painting and Finishing

Exterior: Plywood panel siding, facia, shutters, and soffit areas should be stained with pigmented stain as required. Use a light color stain for the trim and shutters, and a darker stain for the panel siding—or a reverse color selection as desired by owner.

Window and door frames, window sash, screen doors, and similar millwork will have to be painted.

Interior: All woodwork should be painted in semi-gloss. Walls and ceilings are best finished in latex flat, bathroom woodwork and ceiling in semi-gloss.

Termite protection: Termite protection should be provided in termite areas by means of soil treatment or termite shields—or both—as required by local practices and regulations.

Concrete Blocks: Concrete blocks (used as an alternate to foundation of treated wood posts) are laid over concrete footings, as shown on the foundation plan. Footings should be level and laid out to conform to the building line. Blocks are laid up with ⅜-inch-thick mortar joints, and the joints tooled on all exposed exterior surfaces. Anchor straps, when used, are embedded to a depth of at least 12 inches.

BILL OF MATERIALS

The following materials are required for construc-

tion of the home in this plan.

Foundation

The foundation of treated posts set on concrete footings requires:

—1¼ cubic yards concrete

—15 treated foundation posts, 5 feet long or longer as required by slope, with 7-inch minimum top diameter

Floor Framing

Floor framing consisting of floor joists supported on ledgers nailed to the anchored floor beams requires:

—4   2 by 12s, 12 feet long
—12 2 by 12s, 16 feet long
—46 2 by 8s, 12 feet long
—12 2 by 4s, 12 feet long
—40 lineal feet of 22-gauge by 2-inch galvanized anchor strap

Floor

Requirements of floor tile and subfloor are:

—48   4-by-8-foot sheets of 3.4-inch tongue- and-groove plywood
—2,600   9-by-9-inch asphalt tile

Wall and Partition Framing

Framing material for walls and partitions (studs and plates) includes:

—133 2 by 4s, 8 feet long
—144 2 by 4s, 12 feet long
—16 2 by 6s, 12 feet long

Ceiling and Roof Framing

Materials for rafters, joists, and a flush beam over living area are:

—25 2 by 4s, 8 feet long
—6   2 by 4s, 12 feet long
—54 2 by 6s, 20 feet long

—26 2 by 8s, 12 feet long
—25 2 by 8s, 14 feet long
—3 1 by 8s, 12 feet long

Roof

Roofing and sheathing requirements are:

—45 4-by-8-foot sheets of ⅜-inch plywood sheathing grade (standard)
—14 squares of 210-pound asphalt shingles

Siding

The following siding or equivalent alternates are required:

—10 4-by-9-foot sheets of ⅜-inch plywood, exterior grade (textured surface)
—16 1-by-2-inch battens, 12 feet long
—28 4-by-9-foot sheets of rough-textured, ⅝ inch plywood

Windows

All windows are double-hung and purchased treated and complete with screens, and storms when required. Quantity of each size is:

—1 ³⁶/₂₄
—7 ³²/₂₀
—5 ²⁴/₂₀
—1 ²⁴/₁₆

Exterior Doors

Doors are 1/34-inch thick and glazed. Frame, trim, and hardware are required for each. Screen doors are furnished when required. Sizes are:

—Front: 3 feet wide and 6 feet, 8 inches high
—Rear: 2 feet, 8 inches wide and 6 feet, 8 inches high

Insulation

Blanket insulation with aluminum foil vapor barrier on one side is required for ceiling, walls, and floor.

—2,000 square feet, 2 inches thick (or as required), 16 inches wide.
—1,200 square feet, 3 inches thick (or as required), 16 inches wide.

Roof Ventilators

Requirements for ventilating the roof are:

—2 peak-type outlet vents
—16 square feet of screen for inlet vent slots

Frieze Board

Material required for frieze board is:

—5 4-by-8-foot sheets of ⅜-inch plywood, exterior grade, textured surface

Interior Wall and Ceiling Finish

Gypsum board is used for all interior finish except two accent walls in living-dining area, second floor bedroom end walls, and bathroom (except for plastic-coated hardboard above the bathtub). Requirements are:

—74 4-by-8-foot sheets of ⅜-inch gypsum board
—3 4-by-8-foot sheets of plastic-coated hardboard with corner and edge moldings (bath)
—22 4-by-8-foot sheets of ¼ prefinished plywood paneling
—3 250-foot rolls of joint tape
—6 25-pound bags of joint compound

Stairs

Material required for stairs are:

—2 1 by 10s, 14-foot-long stringers
—2 2 by 4s, 10 feet long
—2 2 by 12s, 14 feet long
—3 1 by 8s, 10 feet long
—11 Treads
—16 board feet of 1-by-4-inch fir flooring (platform)

Interior Doors

Interior door requirements are:

—1 set of 2-foot, 4-inch by 6-foot, 8-inch hollow-core door with jambs, stops, and hardware
—2 sets of jambs for 2 foot, 6-inch by 6 foot, 8-inch doors
—2 sets of jambs for 3 foot by 6 foot, 6-inch doors
—1 4-by-8-foot sheet ½-inch plywood, interior AC

Interior Trim

Trim for windows, doors, and base includes:

—330 feet of ⁹⁄₁₆-by-2⅛-inch casing
—340 feet of ½-by-3-inch base

Cabinets

Material requirements for wood-frame and plywood cabinets are:

—3 4-by-8-foot sheets of 3¼-inch plywood, interior AC
—1 1 by 3s, 8 feet long
—6 1 by 4s, 8 feet long

Wardrobes

Wardrobes consisting of closet poles with a shelf over require:

—5 4-by-8-foot sheets of ¾-inch plywood, interior AA
—6 4-by-8-foot sheets of ½-inch plywood, interior AA
—4 closet poles, 1⁵⁄₁₆-inch-diameter by 8 feet long
—2 1 by 3s, 8 feet long
—7 pair of pole sockets for 1⁵⁄₁₆-inch poles

Front Stoop

The front stoop, consisting of planks laid across 2-by-4 framing, supported by treated posts, has the following material requirements:

—2 Treated posts, 5 feet long with 6-inch minimum top diameter.
—3 2 by 4s, 8 feet long
—10 2 by 6s, 6 feet long
—4 ½-inch galvanized carriage bolts, 8 inches long

Rear Stoop

Material requirements for the rear stoop are:

—3 Treated posts, 5 feet long, with 6-inch minimum top diameter.
—4 2 by 4s, 8 feet long
—9 2 by 6s, 8 feet long
—6 ½-inch galvanized carriage bolts, 8 inches long

Nails

The nails needed are:

—52 pounds, eightpenny common
—30 pounds, sixteenpenny common
—3 pounds, twentypenny common
—4 pounds, tenpenny common, galvanized
—2 pounds, sixteenpenny common, galvanized
—3 pounds, fourpenny finish
—20 pounds, eightpenny finish
—18 pounds, 3-inch galvanized roofing
—30 pounds, fourpenny cooler

Paint and Finish

Quantities required for one coat of paint inside, two on the exterior, and one coat of stain outside are:

—11 gallons for walls and ceiling (interior)
—2 gallons for interior trim
—7 gallons for exterior siding and trim (Use different body and trim color, as desired)
—1 gallon for exterior paint (windows, doors, and frames)

Electrical

In addition to rough wiring and 100-amp service, the following items are required:

—17 duplex outlets
—2 wall-mounted interior lights
—2 wall-mounted exterior lights
—3 ceiling lights
—5 switches, single pole
—2 switches, 3-way

Heating

The heating system requires a 100,000 BTU gas furnace with 10 registers, 6 cold-air returns, a duct, a prefab chimney, controls, and other items required for a complete system.

Plumbing

Complete plumbing must be provided for the following required fixtures:

—5-foot bathtub
—water closet
—lavatory
—kitchen sink
—50-gallon water heater
—16-by-16-by-24-inch vitrified tile or equal (at sewer and water entrance)

OPTIONS

Floor

When tongue-and-groove plywood is not available, square-edge plywood with all edges blocked can be used. Material requirements for flooring and subfloor are:

—48   4-by-8-foot sheets of ¾-inch plywood
—20   2 by 4s, 12 feet long
—2,600 9-by-9-inch asphalt tile

Foundation

Instead of a foundation of treated posts on concrete footings, you could have a foundation of concrete blocks on poured concrete footings, plus masonry foot and rear stoops. This requires:

—7   cubic yards of concrete (footings)
—504 8-by-8-by-16-inch concrete blocks

—85   4-by-8-by-16-inch concrete blocks
—72   8-by-16-by-16-inch concrete blocks (based on 4 foot foundation depth.)
—24   8-by-6-by-8-inch concrete blocks (stoop)
—18   sacks of prepared mortar
—2   cubic yards of mason's sand
—60   square feet 6-by-6-inch mesh reinforcing (stoops)
—4   ½-inch reinforcing rods, 5 feet long (step)
—4   ½-inch reinforcing rods, 7 feet long (step)
—800 square feet of 4-mil polyethylene film (soil cover)
—2   8-by-16-inch foundation vents
—1   16-by-16-inch (minimum) access door and frame

Floor Framing

Materials required for a concrete block foundation are:

—4   2 by 12s, 16 feet long
—58   2 by 8s, 12 feet long
—4   2 by 4s, 16 feet long
—45   lineal feet of 22-gauge-by-2-inch anchor strap

Second-Floor Bathroom

A second bathroom can be added in a second-floor dormer. Materials required for constructing the dormer and providing bathroom fixtures are:

—29 2 by 4s, 8 feet long
—14 2 by 4s, 16 feet long
—12 2 by 6s, 8 feet long
—4   4-by-8-foot sheets of ⅝-inch plywood, exterior grade (rough textured)
—16 1 by 2 battens, 8 feet long
—7   4-by-8-foot sheets of ¼-inch prefinished plywood
—2   4-by-8-foot sheets of ⅜-inch gypsum board
—40 9-by-9-inch asphalt tile (with adhesive)
—1   ²⁴⁄₁₆ treated, double-hung window (storm and screen)

—1 set of 2 foot, 4-inch by 6-foot, 6-inch hollow-core door with jambs, stops, and hardware
- —1 wall-mounted interior light
- —1 switch
- —1 hot-air register with duct
- —1 30-by-30-inch shower stall
- —1 lavatory
- —1 water closet

Note: As a less costly addition, a lavatory could be installed in each upstairs bedroom.)

Bedroom Dormer

A bedroom dormer can be included with the second-floor bathroom addition. Materials required are:

- —20 2 by 4s, 8 feet long
- —12 2 by 4s, 16 feet long
- —2 4-by-8-foot sheets of $\frac{5}{8}$ inch plywood, exterior grade (textured surface)
- —6 1 by 2 battens, 8 feet long
- —6 4-by-8-foot sheets of $\frac{3}{8}''$ gypsum board
- —56 9-by-9-inch asphalt tile (with adhesive)
- —1 $^{24}/_{16}$ treated double-hung window (storm and screen)

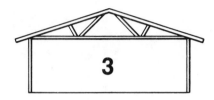

3

Construction of First-and
Second-Generation Homes

The basic building materials are wood and the many wood-base materials now on the market. The same basic materials used in the most expensive homes are also used in low-cost wood homes. Improved housing for today's families means those that are low in cost, easily maintained, and equipped for good family living—but, attractive as well.

Whether you plan to build the entire house yourself or hire a contractor for specific portions, you will want to learn about the many phases of building. Even if you decide that you don't want to do any of the work yourself, you will still need the knowledge in order to speak intelligently, to understand fully what a contractor says, and to make sure your plans are carried through as you want them.

You will have many decisions to make. Is this going to be a first-generation, low-cost house or one that you will call home for many years to come? One of the major causes of dissatisfaction with first-generation housing is that the family's needs constantly change. These changes are due to an increase or decrease in family size, a change in economic status and social activities, or a change in the desires of family members.

This chapter will provide a step-by-step information guide on building both permanent-type residences for the settled, and low-cost houses for first-generation home owners.

FIRST-GENERATION HOME

This modular construction design provides a house with the flexibility to allow major changes in floor plans without major structural change; allows home repairing that approaches the simplicity of replacing TV tubes; provides a house that can be mass-produced without excessive uniformity. This house makes maximum use of economic and time-saving equipment and procedures.

Approximately 50 percent of this house is built of standard premanufactured parts, including the footings, foundation panels, roof trusses, exterior wall panels, and the interior partitions. Complete plans for this house are shown in Figs. 3-1 and 3-2.

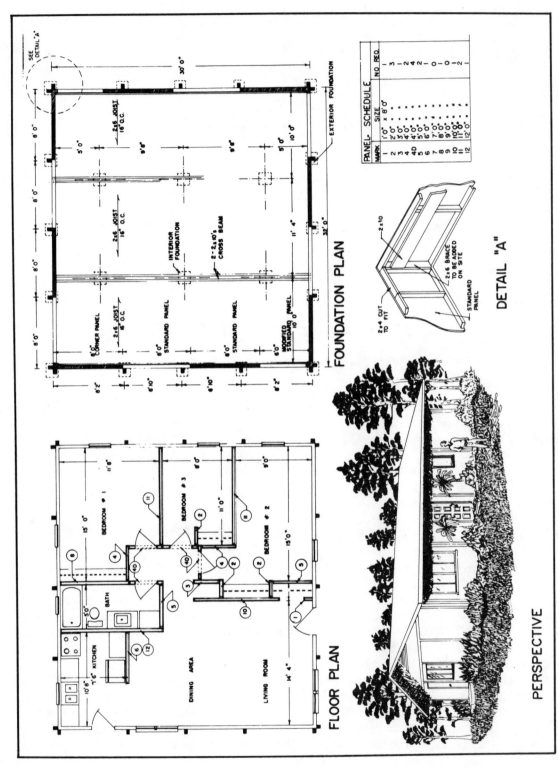

FOUNDATION PLAN

PANEL SCHEDULE		
MARK	SIZE	NO. REQ.
1	1'-0" x 8'-0"	3
2	2'-0" "	—
3	3'-0" "	2
4	4'-0" "	4
4D	4'-0" "	2
5	5'-0" "	—
6	6'-0" "	—
7	7'-0" "	0
8	8'-0" "	0
9	9'-0" "	0
10	10'-0" "	2
11	11'-0" "	
12	12'-0" "	

DETAIL "A"

FLOOR PLAN

PERSPECTIVE

Fig. 3-1. Perspective, floor plan, and foundation plan for first-generation house. Courtesy USDA

46

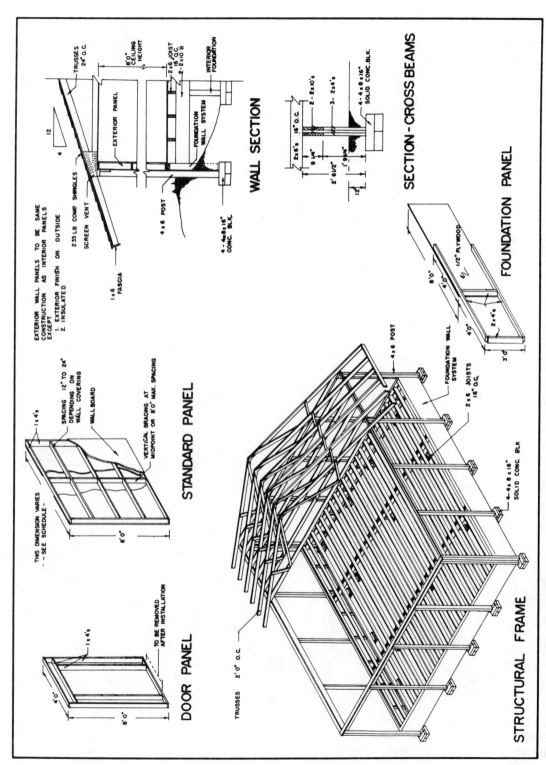

STANDARD PANEL

THIS DIMENSION VARIES
— SEE SCHEDULE —

1 x 4's

SPACING 12" TO 24"
DEPENDING ON
WALL COVERING

WALLBOARD

VERTICAL BRACING AT
MIDPOINT OR 8'0" MAX. SPACING

8'0"

DOOR PANEL

1 x 4's

TO BE REMOVED
AFTER INSTALLATION

4'0"

8'0"

WALL SECTION

TRUSSES
24" O.C.

8'0"
CEILING
HEIGHT

2 x 6 JOIST
16" O.C.
2 - 2 x 10 B

INTERIOR
FOUNDATION

EXTERIOR PANEL

FOUNDATION
WALL SYSTEM

12

4

EXTERIOR WALL PANELS TO BE SAME
CONSTRUCTION AS INTERIOR PANELS
EXCEPT
1. EXTERIOR FINISH ON OUTSIDE
2. INSULATED

235 LB COMP. SHINGLES

SCREEN VENT

1 x 6
FASCIA

4 x 6 POST

4 - 4 x 8 x 16"
CONC. BLK.

SECTION - CROSS BEAMS

2 - 2 x 10's

3 - 2 x 4's

4 - 4 x 8 x 16"
SOLID CONC. BLK.

16" O.C.

2 x 6's

9 1/4"

9 1/4"

2' 6 1/2"

1' 9 1/4"

12"

FOUNDATION PANEL

1/2" PLYWOOD

2 x 4's

8'0"

4'0"

4'0"

3'0"

STRUCTURAL FRAME

4 x 6 POST

FOUNDATION WALL
SYSTEM

2 x 6 JOISTS
16" O.C.

4 - 4 x 8 x 16"
SOLID CONC. BLK

TRUSSES 2'0" O.C.

Fig. 3-2. Structural frame, exterior-wall section, cross-beam section, and various panels for first-generation house. Courtesy USDA
USDA

47

Foundation and Framing

Footing: A continuous ditch around the perimeter of the house (Fig. 3-3) provides a base for the pole footings, which are easily leveled with a simple 8-foot level. The footings are solid concrete blocks, set at the bottom of the ditch and leveled in a bed of mortar. Crushed rock added to the ditch forms a perimeter drain.

Foundation Panels: The foundation panels are 8-foot modules. Each module consists of a pressure-treated, ½-inch plywood skin and an 8-foot frame of 2 by 4s that supports half of the attached plywood skin and projects to support half of the skin of the adjacent panel (Fig. 3-4). This frame joins the modules and provides structural continuity from panel to panel. The panels, which are fastened together and properly aligned on the leveled footings, form a continuous wall around the structures.

Fig. 3-4. Modules of pressed foundation panels.

The concrete footings and panels are installed with little difficulty, unless panels are not square. Panels with flaws project irregularities to other panels.

Pole Frame

The pole frame (which could be precut at the factory) is assembled on the ground at the building site and tilted into place on the footings in three-to-five-pole sections (Fig. 3-5). After each section is tilted into place, the sections are joined together, and a plate of 2-by-8-inch lumber is attached to the perimeter beam to stiffen and strengthen the walls. The pole frame is custom-fitted and fastened to each pole.

Fig. 3-3. Ditch around the perimeter of the house.

Fig. 3-5. Framing tilted into place on site.

Fig. 3-6. Floor joists.

Fig. 3-7. Roof trusses.

Floor

The ½-inch plywood floor is supported on 2-by-6-inch floor joists placed 16 inches on center (Fig. 3-6). The 2 by 6s are continuous for the entire length of the house. The floor joists are supported by four main beams that extend across the width of the house. These main beams, built of 2 by 10s, are also structurally spliced to be continuous. The two end beams are supported on the foundation wall, and the two center beams are supported on posts. The posts are custom-fitted to hold the main beams level. Joints between adjacent sheets of plywood are supported by 2-by-4-inch splicing blocks, which are placed between the floor joists and nailed in place to prevent differential deflection between plywood sheets. The entire floor is custom-built on site.

Roof Trusses

The roof trusses are constructed in a simple homemade jig (Fig. 3-7). Each truss is carried into the structure upside down, set on the perimeter plate beam, and tilted into an upright position. Next, the truss is moved into place along the lengths of the structure and adjusted for proper overhang on each side. The truss is toenailed in place and then tied down with plumber's strap. The ends of all trusses are marked with a chalk line and cut off square. A 2 by 4 across the ends of the trusses maintain uniform spacing 2 feet on center, supporting the end

of the roof, and providing a solid base for attaching gutters and trim.

Roof

The roof of the house illustrated here is installed in the conventional manner with ½-inch plywood and standard 235-pound, three-tab shingles. A power nailer is used for fastening the plywood in place, but the shingles are hand-nailed.

With the roof installed, the complete frame is a clear-span structure with pole supports around the perimeter, at approximately 8-foot intervals.

Ceiling

The entire ceiling is installed (Fig. 3-8) on the bottom chords of the trusses before any panels or other

Fig. 3-8. Installation of ceiling and carpet before interior wall placement.

materials are carried into the house, thus allowing full sheets of drywall to be used with a minimum of cutting, fitting, and framing. Blocking, framing, and fitting are required around the perimeter of each room when individual rooms have ceilings installed separately.

Three basic exterior walls are built and installed in this house. The first wall is an extruded polyurethane panel 9 feet tall and 4 inches thick, which is surfaced with panel board on the interior and heavy kraft paper on the exterior (Fig. 3-9). Any length of panel was available, but the panels in this house are 8 or 16 feet long. The wall is extruded and it contains 2-by-4-inch aluminum studs (8 feet long) at 2-foot intervals. The second wall panel is frame built of 2 by 4s placed horizontally and spaced vertically

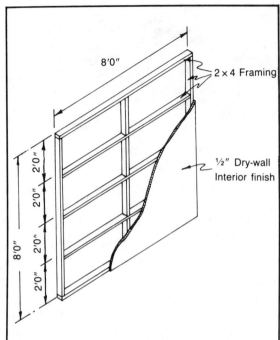

Fig. 3-10. Framed wall panel with horizontal 2 by 4s spaced 2 feet on center. Courtesy USDA

at 2-foot intervals (Fig. 3-10). The interior finish of ½-inch drywall was attached to the frame at the time of construction, but the exterior covering and insulation were omitted until after the walls were in place. The third wall panel is very similar to the second, except the 2 by 4s are vertical studs.

All exterior panels are ¼ inch shorter than ceiling height and are fitted into place and secured with wedges, which forces them against the ceiling. The panels are attached to the pole frame with two 8-inch lag bolts at each pole. A special molding was required to cover and hide the undesirable appearance of the lag bolts.

The panels were finished on only one side before being set in place. They were nailed to the floor and ceiling and thus cannot be easily moved.

Interior Partitions

The interior partitions in this model are constructed of standard panels built in 1-foot lengths up to 12 feet (Fig. 3-11). A special 4-foot door panel was built for a standard 3-foot door. All panel frames are con-

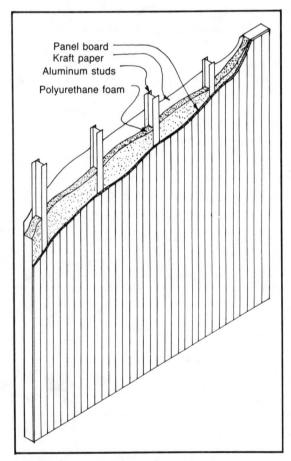

Fig. 3-9. Extruded polyurethane panel with 2-by-4-inch aluminum studs. Courtesy USDA

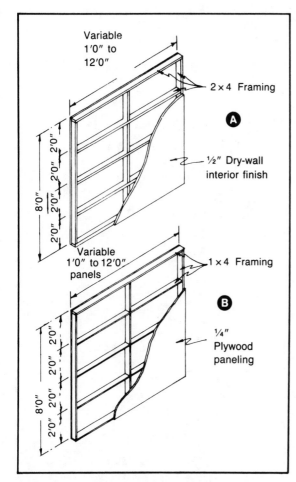

Fig. 3-11. Original and redesigned interior wall panels. A: Original panel with 2-by-4-inch framing and ½-inch drywall finish. B: Redesigned panel with 1-by-4-inch framing and ¼-inch plywood finish. Courtesy USDA

structed of 2 by 4s (similar to the second exterior wall panels) with drywall finish on both sides (Fig. 3-11A). Because of their excessive weight, these partitions were dismantled, and new panels were built with 1 by 4s, using the same horizontal framing pattern (Fig. 3-11B).

To further reduce weight, ¼-inch plywood paneling was used as the finished surface on both sides. Before any partitions were installed, all carpet, pads, and floor tile were installed. Interior partition panels were built ¼ inch shorter than ceiling height. The partition panels were tilted into place, forming the walls for all rooms and closets. Each

panel was held securely by friction between the panel and the ceiling by wedges placed under the panels. Most walls also received some lateral support from other panels placed at right angles to them. End caps were custom-fitted over all exposed ends of panels after they were in place.

Doors and Windows

In this model no special framing was required for the installation of doors and windows because all exterior walls are non-load-bearing curtain walls. Proper-size holes were cut in the polyurethane panels, and the window units were inserted into the openings. The interior trim was then attached to secure the window units and to decorate the perimeter of the openings (Fig. 3-12). The doors and frames were installed in a similar fashion.

Fig. 3-12. Inside wall.

Utilities

The wiring in this house is a surface-mounted complex contained in a two-piece steel baseboard ½ by 2 inches. A prewired harness was installed in prestamped outlet openings at 30-inch intervals. The feeder wires enter the house through a standard custom-wired, 200-amp service entrance equipped with circuit breakers.

The standard custom-installed plumbing system is supplied with city water, and the sewage system is connected to a central aeration plant.

SECOND-GENERATION HOME

The first-generation house was built from standardized panels. Here, a second-generation house—designed by the USDA, focuses on several improvements that could be made.

Foundation and Framing

Foundation Panels: Several improvements were made in the basic foundation panel. The new foundation panel has a full-perimeter frame. It supports the entire plywood skin and features 1-by-4-inch double plates with staggered joints that overlap, with the adjacent panel forming a telescop-

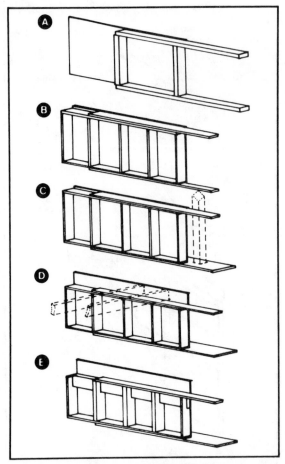

Fig. 3-14. Progressive development in foundation panel design. A: First generation panel with 2 by 4-inch framing and ½-inch plywood. B: Panel with redesigned frame of 1 by 4s for full skin support and for overlap of framing members. C: Panel with wider bearing plate for supporting poles and eliminating concrete footings. D: Panel with dropped top plate for providing slot for floor joists. E: Panel with 2 by 6-inch top bearing plate for carrying roof and floor loads. Courtesy USDA

ing joint (Fig. 3-13). Figure 3-14 shows several panel modifications that can be incorporated for special uses.

The redesigned foundation is made up of six standard panels that can be assembled to fit houses of almost any shape. Straight panels are built in 2-, 4-, 6-, and 8-foot lengths (Fig. 3-15). A standard corner panel and an outside or obtuse corner panel reduce on-site cutting and fitting and improve the structural stability of the corner joint (Fig. 3-16). The

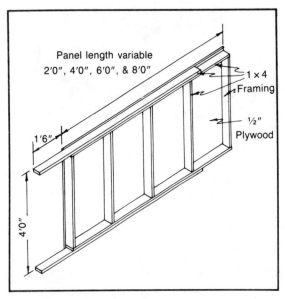

Panel length variable
2'0", 4'0", 6'0", & 8'0"

1 × 4 Framing

½" Plywood

1'6"

4'0"

Fig. 3-13. Redesigned foundation panel. Courtesy USDA

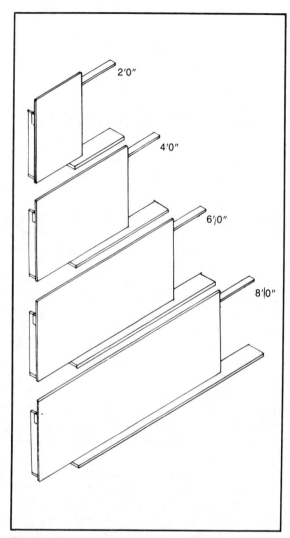

Fig. 3-15. Redesigned foundation panels built in four different lengths. Courtesy USDA

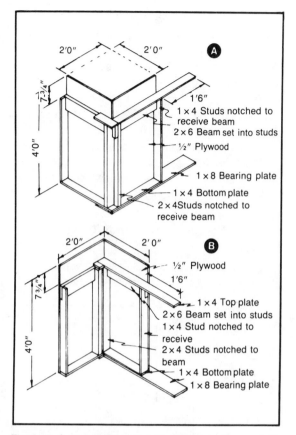

Fig. 3-16. Corner panel design. A: Obtuse corner panel. B: Standard corner panel. Courtesy USDA

two corner panels allow the fitting of standard panels with other than rectangular floor plans. All panels are joined by a telescoping connection, which allows a panel to slide into its precise location. The telescoping joint can be expanded up to 10 inches for structures that are not modular in length (Fig. 3-17).

When joints are expanded or telescoped, a spacer block must be added to cover the space between panels. The spacer blocks must be made from pressure-treated wood of the same quality used in the basic foundation panels.

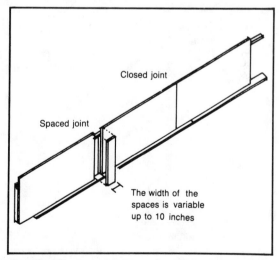

Fig. 3-17. Standard foundation panel with telescoping joints. Courtesy USDA

Pole Frame

The problem of water flowing into the crack at the base of the wall panels was solved by placing a 1-inch filler strip between the poles and the foundation panels (Fig. 3-18).

The poles of the pole frame were lengthened approximately 8 inches to raise the perimeter beam above the ceiling line and provide full clearance for inserting modular units into the space between the floor and the ceiling. This change created the need to redesign the roof truss in order to maintain the ceiling at its original height.

Dropped-Chord Roof Truss

The extension of the post and the intrusion of the beam resulted in a new truss design that is superior in many aspects to the truss outlined in first-

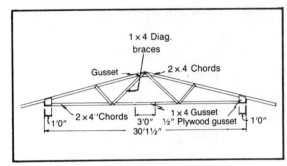

Fig. 3-19. Dropped-chord roof truss. Courtesy USDA

generation house. The new truss, or "dropped-chord truss," was so named because the lower tension chord was lowered 8 inches to a position between the perimeter beams, thus dropping the ceiling below the perimeter beams (Fig. 3-19).

The interior braces were lengthened, and larger gusset plates were required to assemble the truss. Because the dropped chord destroyed the regular triangular sections of the truss, it was necessary to conduct loading tests on the new truss, using the same nailing schedule as designed for a regular 30-foot truss at a 25-pound load.

All trusses that failed did so as a result of local bending of the top chord between the end gusset and first brace (Fig. 3-20). In no case did the interior braces fall, nor were there any indication of failures at the nailed points.

The dropped-chord truss provides several structural and construction advantages. First the bottom chord of the truss fits between the perimeter beams instead of on top of them (Fig. 3-21), thus providing a positive, uniform spacing of the side walls of the house along its entire length.

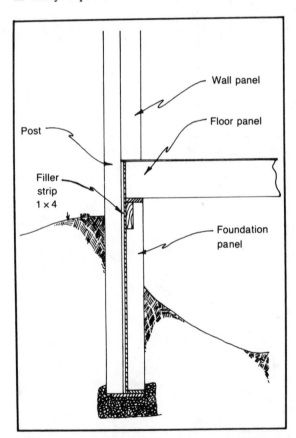

Fig. 3-18. Foundation wall section with filler strip. Courtesy USDA

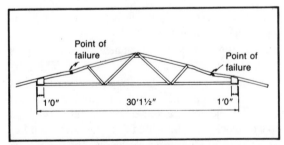

Fig. 3-20. Roof truss showing points of failure under a uniformly distributed load of 40 pounds per square foot. Courtesy USDA

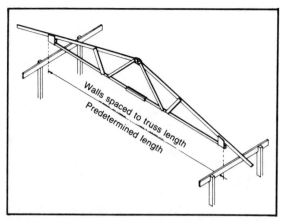

Fig. 3-21. Roof-truss and pole-frame assembly. Courtesy USDA

Second, the truss no longer needs to be fastened in place by toenails and straps, since nails are driven through the perimeter beam and into the end of the bottom chord of the truss, providing a rigid, shear-type joint.

Third, the dropped chord provides space for placement of insulation (Fig. 3-22) and eliminates one of the common attic ventilation problems: insulation spilling into or being stuffed into the overhanging section of the roof and inhibiting airflow, which creates a sweating and decaying problem.

Split-Frame Wall Panels

The exterior walls for the second-generation house are split-frame panels (Fig. 3-23). With the split-frame concept, a wall panel is finished on only one side. To form a complete wall, interior and exterior panels are required, and two rows of panels are placed back to back. Because the two rows are independent, the interior panels can be altered, changed, removed, or replaced without disturbing the exterior panels. The thickness of the wall can be varied by shifting the interior row of panels (Fig. 3-24).

A need for added insulation might require an increase in wall thickness. Insulation can be installed

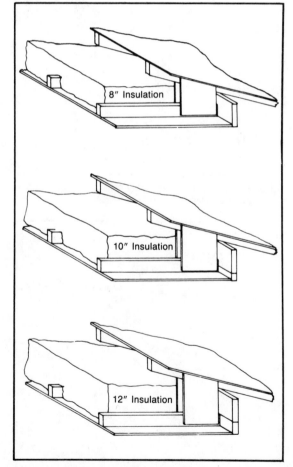

Fig. 3-22. Insulation pockets created by dropped chord. Courtesy USDA

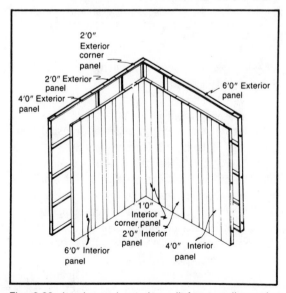

Fig. 3-23. Interior and exterior split-frame wall panels. Courtesy USDA

55

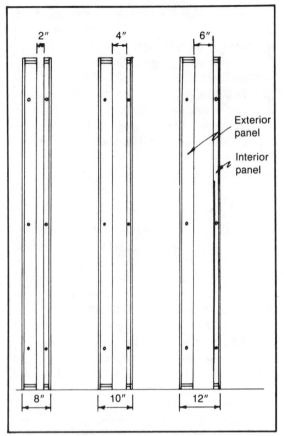

Fig. 3-24. Cross-section of exterior wall showing variations in panel spacing for adding insulation or for hiding obstructions. Courtesy USDA

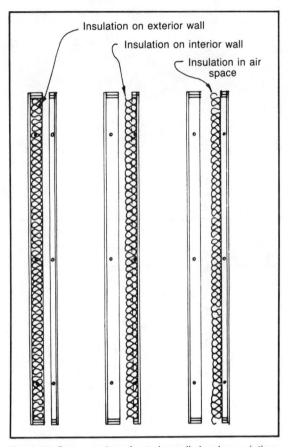

Fig. 3-25. Cross-section of exterior wall showing variations in insulation placement. Courtesy USDA

on exterior or interior panels or in the space between the panels (Fig. 3-25).

The exterior panels have a ½-inch exterior plywood subsiding mounted on a frame of 1-by-4s; a better grade of finish can be added after installation. The panel is built in 1-foot increments—up to 8 feet—and special filler strips are designed in 1-to-11-inch lengths to complete the length of irregular or non-modular spans (Fig. 3-26). Corners are formed by special corner panels that eliminate cutting and fitting and provide a strong, tightly sealed structural and functional unit for either standard or obtuse corners (Fig. 3-27).

The windows and doors are contained in special panels. Door or window panels are inserted in the place of standard 4- or 6-foot panels (Fig. 3-28).

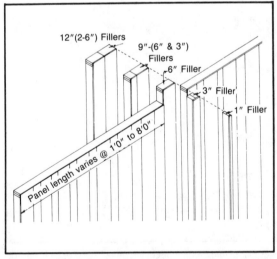

Fig. 3-26. Exterior panels and filler strips. Courtesy USDA

56

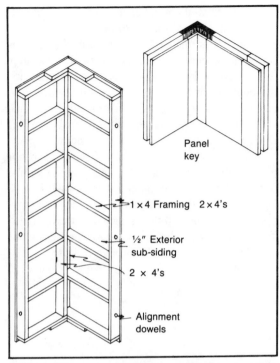

Fig. 3-27. Exterior corner split-frame panel. Courtesy USDA

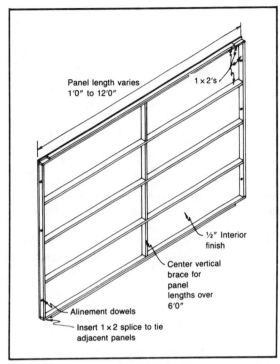

Fig. 3-29. Interior panel of exterior wall. Courtesy USDA

The door and window panels are not split framed, but they have a subsiding exterior finish and an interior finish.

The interior finish of the interior panel is attached to a frame of 1 by 2s. The panels are avail-

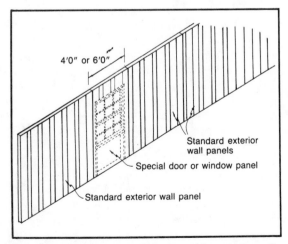

Fig. 3-28. Special door or window panel for exterior walls. Courtesy USDA

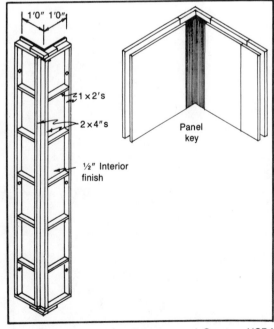

Fig. 3-30. Interior corner split-frame panel. Courtesy USDA

57

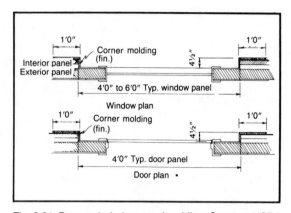

Fig. 3-31. Door and window-panel molding. Courtesy USDA

able in 1-to-12-foot lengths (Fig. 3-29). The basic frame is similar to that of the exterior panels, and the prefabricated corner panels reduce on-site fitting to a minimum (Fig. 3-30).

The interior panels form the finished wall spans between the door and window panels. Each span of wall is independent of other spans, so a starting and finishing molding or paneling is used at each door and window panel (Fig. 3-31). This molding or paneling includes a cover place that extends 12 inches along the wall on each side of the window or door, and serves as the interior finish from the interior panel to the finished window. Thus, 1-foot modular panels will not fit or complete a span of wall; 1-to-12 inch nondecorative filler panels are added to complete the span behind the 12-inch decorative cover. The complete plans for the second-generation house are shown in Figs. 3-32, through 3-34.

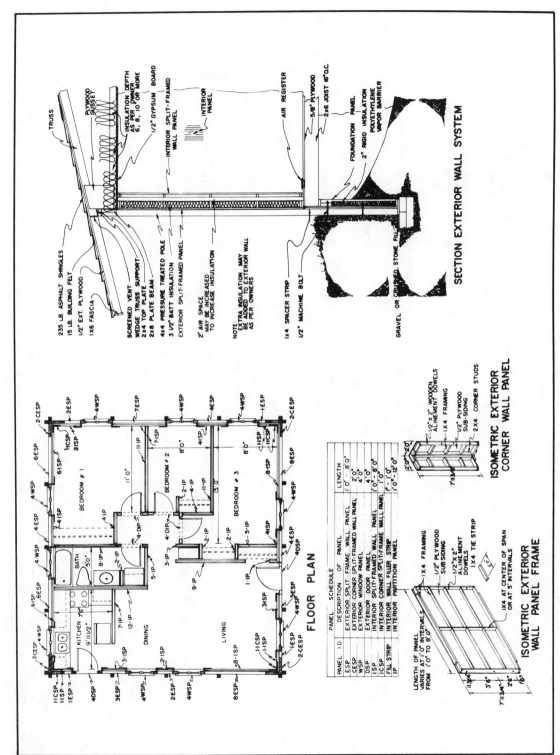

SECTION EXTERIOR WALL SYSTEM

235 LB. ASPHALT SHINGLES
15 LB. BUILDING FELT
1/2" EXT. PLYWOOD
1X6 FASCIA

TRUSS
PLYWOOD GUSSET
INSULATION DEPTH AS PER OWNER 6, 8, 10 OR MORE
1/2" GYPSUM BOARD
INTERIOR SPLIT-FRAMED WALL PANEL
INTERIOR PANEL
AIR REGISTER

SCREENED VENT
WEDGE TRUSS SUPPORT
2X4 TOP PLATE
2X8 PLATE BEAM
4X4 PRESSURE TREATED POLE
3 1/2" BATT INSULATION
EXTERIOR SPLIT-FRAMED PANEL

2" AIR SPACE MAY BE INCREASED TO INCREASE INSULATION

NOTE: EXTRA INSULATION MAY BE ADDED TO EXTERIOR WALL AS PER OWNERS

1X4 SPACER STRIP
1/2" MACHINE BOLT

FOUNDATION PANEL
5/8" PLYWOOD
2X6 JOIST 16"O.C.
2" RIGID INSULATION
POLYETHYLENE VAPOR BARRIER

GRAVEL OR CRUSHED STONE FILL

FLOOR PLAN

BEDROOM #1
BEDROOM #2
BEDROOM #3
KITCHEN
DINING
BATH
LIVING

Panel Schedule

PANEL I.D.	DESCRIPTION OF PANEL	LENGTH
ESP	EXTERIOR SPLIT-FRAME WALL PANEL	1'0" - 8'0"
CESP	EXTERIOR CORNER SPLIT-FRAMED WALL PANEL	2'0"
WSP	EXTERIOR WINDOW PANEL	4'0"
DSP	EXTERIOR DOOR PANEL	4'0" - 8'0"
ISP	INTERIOR SPLIT-FRAMED WALL PANEL	1'0"
ICSP	INTERIOR CORNER SPLIT-FRAME WALL PANEL	1'- 8'0"
FILL STRIP	INTERIOR WALL FILLER STRIP	
IP	INTERIOR PARTITION PANEL	1'0"-12'0"

ISOMETRIC EXTERIOR WALL PANEL FRAME

1X4 FRAMING
1/2" PLYWOOD SUB SIDING
1/2"X2" ALINEMENT DOWELS
1X4 TIE STRIP
1X4 AT CENTER OF SPAN OR AT 5' INTERVALS

LENGTH OF PANEL VARIES AT 1'0" INTERVALS FROM 1'0" TO 8'0"

ISOMETRIC EXTERIOR CORNER WALL PANEL

1/2"X 2" WOODEN ALINEMENT DOWELS
1X4 FRAMING
1/2" PLYWOOD SUB-SIDING
2X4 CORNER STUDS

Fig. 3-32. Floor plan, exterior wall section, and exterior wall panels for second-generation house. Courtesy USDA

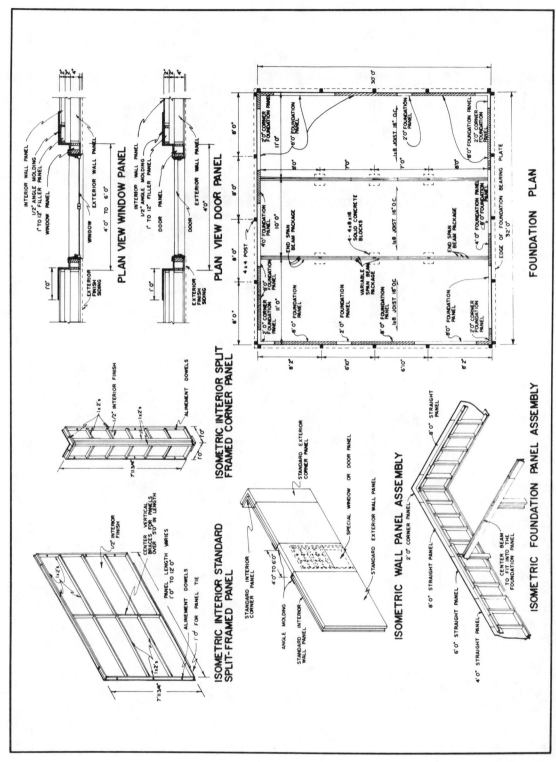

Fig. 3-33. Foundation plan and various panels for second-generation house. Courtesy USDA

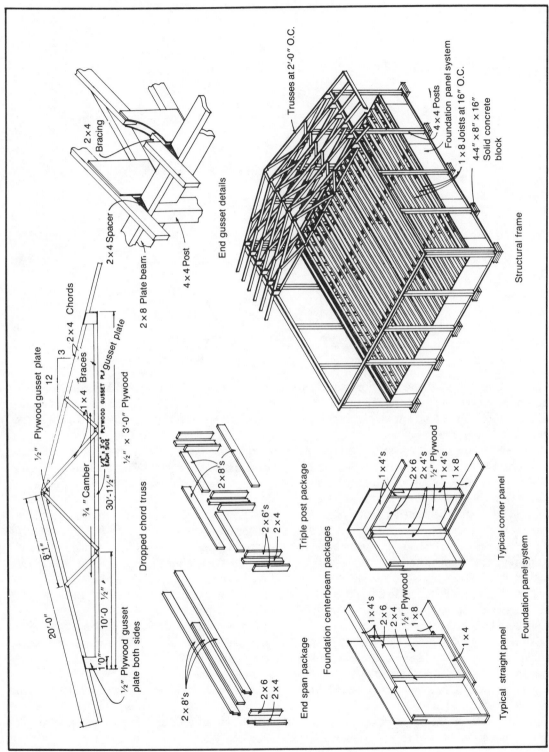

2 × 4 Bracing

End gusset details

2 × 4 Spacer

2 × 8 Plate beam

4 × 4 Post

Trusses at 2'-0" O.C.

4 × 4 Posts

Foundation panel system

1 × 8 Joists at 16" O.C.

4-4" × 8" × 16" Solid concrete block

Structural frame

½" Plywood gusset plate

2 × 4 Chords

3

12

1 × 4 Braces

½" × 3'-0" Plywood gusset plate

½" × 3'-0" PLYWOOD GUSSET PL EACH SIDE

¾" Camber

30'-1½"

Dropped chord truss

½" × 3'-0" Plywood

2 × 8's

Triple post package

2 × 6's

2 × 4

Foundation centerbeam packages

1 × 4's

2 × 6

2 × 4's

½" Plywood

1 × 4's

1 × 8

Typical corner panel

20'-0"

8'1"

10'-0 ½" ±

1'-0"

½" Plywood gusset plate both sides

2 × 8's

End span package

2 × 6

2 × 4

1 × 4's

2 × 6

2 × 4

½" Plywood

1 × 8

Typical straight panel

1 × 4

Foundation panel system

Fig. 3-34. Structural frame, dropped-chord truss, end gusset, and foundation panels and packes for second-generation house. Courtesy USDA

61

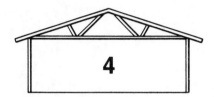

4

Exterior Construction

The floor system, interior and exterior walls, and the roof are the major components of a single-story house. Houses with flat or low-sloped roofs are usually variations of these systems.

Figure 4-1 shows a floor system constructed over a crawlspace area. Supporting beams are fastened to treated posts embedded in the soil or to masonry piers. In south, central, and coastal areas, provisions must be made for protection from termites. Construction of this type of support for the floor joist has a great advantage because grading is not required, so it can be used on relatively steep or uneven slopes. Floor joists are fastened to these beams and the subfloor nailed to the joists. This results in a level, sturdy platform upon which the rest of the house is constructed.

Exterior walls, often assembled flat on the subfloor and raised in ''tilt-up'' fashion, are fastened to the perimeter of the floor platform. Exterior coverings and window and door units are included after walls are plumbed and braced.

Interior walls are usually the next components to be erected unless trussed rafters (roof trusses)

are used. Trussed rafters are designed to span from one exterior sidewall to the other and do not require support from interior partitions. This allows partitions to be placed as required for room dividers. When ceiling joists and rafters are used, a bearing partition near the center of the width is necessary.

Several systems can be used to provide a roof over the house. One consists of normal ceiling joists and rafters, which require some type of load-bearing wall between the sidewalls (Fig. 4-2A). Another is the trussed rafter system (Fig. 4-1 and Fig. 4-2B).

After the site for the house has been selected, all plant growth and sod should be removed. The area can then be leveled for staking and location of the support posts or piers.

The foundation plan in the working drawings for the house shows all the measurements necessary for construction (Fig. 4-3). The first step is to establish a baseline along one side with heavy cord and solidly driven stakes located well outside the end building lines (stakes 1 and 2). This baseline should be at the outer faces of the posts, piers, or foundation walls. When a post foundation is used with an

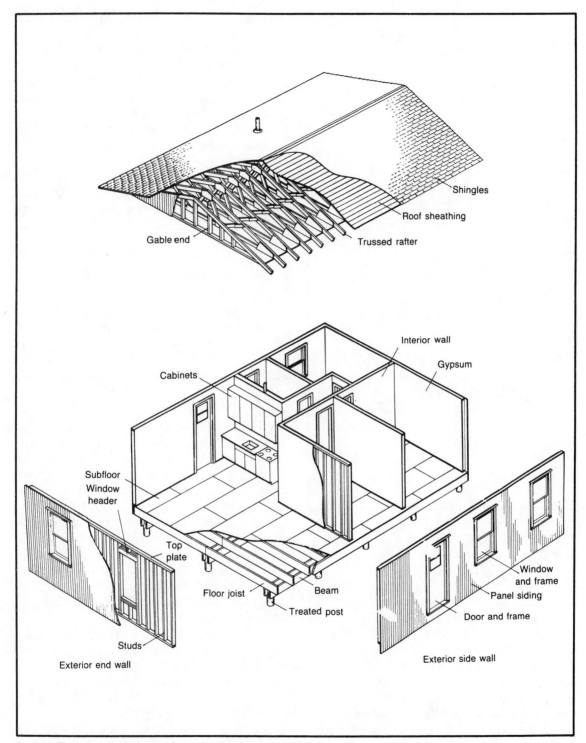

Fig. 4-1. Exploded view of wood-frame house. Courtesy USDA

Shingles

Roof sheathing

Trussed rafter

Gable end

Interior wall

Gypsum

Cabinets

Subfloor
Window
header

Top
plate

Floor joist

Beam

Treated post

Window
and frame

Panel siding

Door and frame

Studs

Exterior end wall

Exterior side wall

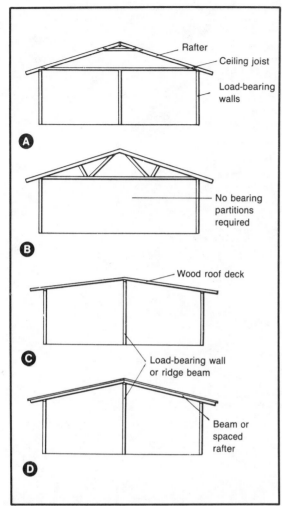

Fig. 4-2. Types of roof construction. A: Rafter and ceiling joists for sloped roofs. B: Trussed rafter for sloped roofs. C: Wood roof decking for low-sloped roofs. D: Beam with wood or fiberboard decking. Courtesy USDA

overhang, the post faces will be 13½ inches in from the building line when a 12-inch overhang is used.

When masonry piers or wood posts are located at the edge of the foundation, the outer faces are the same as the building line. These details are normally included in the working drawings. A second set of stakes (stakes 3 and 5), is established parallel to stakes 1 and 2. When measuring across, be sure that the tape is at right angles to the first baseline. Just as with the 1-2 baseline, this line will locate the outer edge of posts or piers. A third set

of stakes (stakes 5 and 6) is established at one end of the building line.

A square 90-degree corner can be established by laying out a distance of 12 feet at line 1-2 and 9 feet along line 5-6. This design requires no load-bearing walls between the sidewalls. A third design consists of thick wood roof decking (Fig. 4-2C). A fourth is open beams and decking that span between the exterior walls and a center wall or ridge beam (Fig. 4-2D). The truss and the conventional joist-and-rafter construction require some type of finish for the ceiling. The decking (Fig. 4-2C) or the beam and decking (Fig. 4-2D) combinations can serve both as interior finish and as a surface to apply the roofing material.

FOUNDATION SYSTEMS

The first essentials in house construction is to select the most desirable site. A lot in a smaller city or community presents few problems. The front and side yard distances are controlled by local regulations. If the site is in a rural or outlying area, care must be taken in staking out the house location.

Site Selection and Layout

Good drainage is another essential. Be certain that natural drainage is away from the house or that such drainage can easily be assured by modification of ground slope. Avoid low areas. Check that soil conditions are favorable for excavation and for the treated posts or masonry piers of the foundation. Large rocks or other obstructions might require changes in the type of footings or foundation. Short cords can be tied to the lines to mark these two locations. Now measure between the two marks and when the diagonal measurement is 15 feet, the two corners at stakes 1-5 and 3-6 are square. Both diagonals should be the same length.

In some areas of the country, building regulations might restrict the use of treated wood foundation posts. A masonry foundation fully enclosing the crawl space might be necessary or preferred.

Footings

The holes for the post or masonry pier footings can now be excavated to a depth of about 4 feet—or as required by the depth of the frostline. They should be spaced as shown in Fig. 4-3. The depth should be such that the soil pressure keeps the post in place.

Place the dirt a good distance away from the holes to prevent it from falling back in. Size of the holes for the wood post and the masonry piers should be large enough for the footings. When posts or piers are spaced 8 feet apart in one direction and 12 feet in the other, a 20-by-20-inch or 24-inch diameter footing is normally sufficient. In softer soils or if greater spacing is used, a 24-by-24-inch footing might be required.

Posts alone without footings of any type, but with good embedment, are being used for pole-type buildings. However, because the area of the bottom end of the pole against the soil determines its load capacity, this method is not normally recommended in the construction of a house where uneven settling could cause problems. A small amount of settling in a pole warehouse or barn would not be serious. Where soil capacities are very high and posts are spaced closely, it is likely that a footing support under the end of the post would not be required. In a house, however a good, stable foundation is important and adequate footings of some type must be considered.

After the holes have been dug to the recommended depth and cleared of loose dirt, an 8-inch-thick or thicker concrete footing is poured (Fig. 4-4). If premixed concrete is not available, it can be fixed by hand or by a small on-the-job mixer. (Most towns have a rental appliance store where you can rent a small mixer by the hour or day.) Tops of the foot-

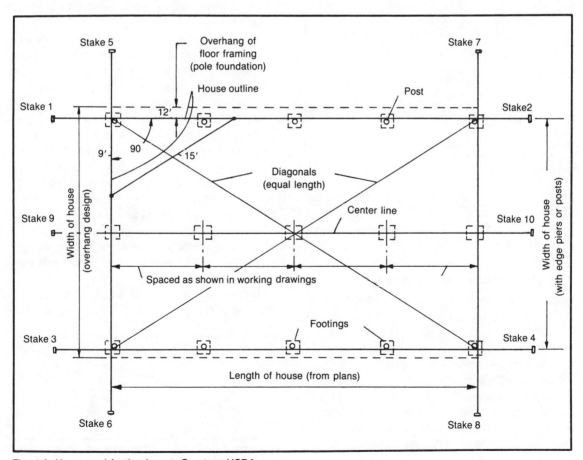

Fig. 4-3. House and footing layout. Courtesy USDA

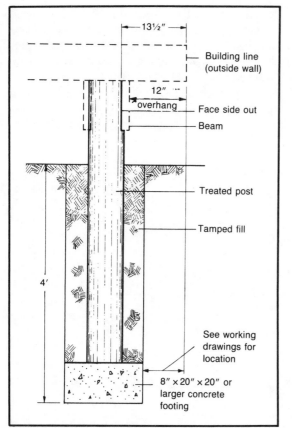

Fig. 4-4. Post embedment and footing alignment; overhang design. Courtesy USDA

ings can be leveled by measuring down a constant distance from the level line. A 5-or 6-bag mix of premixed concrete or a 1: 2½: 3½ (cement: sand: gravel) job-mixed concrete should be satisfactory.

Post Foundations With Side Overhang

Treatment of the foundation posts should conform to Federal Specification TT-W-571. (This may be obtained from the U.S. Department of Agriculture, Washington, D.C.) Penetration of the preservative for foundation posts should be equal to one-half the radius and not less than 90 percent of the sapwood thickness. Select posts according to final finish and appearance. When cleanliness, freedom from odor, or paintability is essential, use preservative-treated posts. Don't use untreated posts in contact with the soil.

For each of the footing locations, select treated posts having a top diameter as shown on the plans. The length can be determined by setting the layout strings to the level of the top of the beams and posts. Select the corner with the highest ground elevation and move the string on the stake to about 18 or 20 inches above the ground level at this point. The minimum clearance under joists or beams should be 12 inches.

Using lightweight string or line level (Fig. 4-5), adjust the cord on the other stakes so that the layout strings around the edge of the building and down the center are all truly level and horizontal. To ensure accuracy, the line must be tight with no sag, with the level located at the center. If available, a surveyor's level works even better. The distance from the strip to the top of concrete footings will now be the length of the posts needed at each location.

A manometer level can also be used in establishing a constant elevation for the posts. This type of level consists of a long, clear plastic tube partly filled with water or other liquid. The water level at each end establishes the correct elevation.

If pressure-treated posts are not available in the lengths just determined, use poles more than twice as long as required. Saw them in half and use with the treated end down. With a saw and a hand ax or drawknife, slightly notch and face one side for a distance equal to the depth of the beams (Fig. 4-6A). The four largest diameter posts should be used for the corners and notched on two adjacent sides (Fig. 4-6B). Facing should be about 1½-inch wide, except for corners or when beam joints might be made and 2½-inch is preferable.

Treated 6-by-6- or 8-by-8-inch posts can be used in place of the round posts when available. Al-

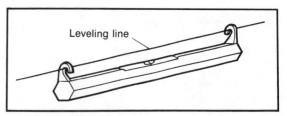

Fig. 4-5. Locate line level midway between building corners when leveling. Courtesy USDA

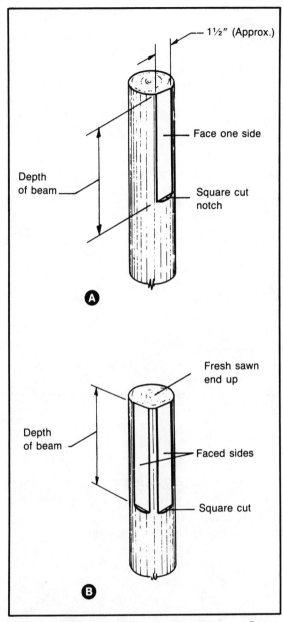

Fig. 4-6. Facing posts. A: Side or intermediate post. B: corner post (use largest). Courtesy USDA

though they might cost somewhat more and do not have the resistance of treated round posts, square posts will reduce on-site labor time.

Locate the notched and faced posts on the concrete footings using the cord on the stakes as a guide

for the faced sides. Place and tamp 8 to 10 inches of dirt around them initially to hold them in place. Posts should be vertical and the faced side aligned with the cord from the stakes along each side, the center, and the ends of the house outline. When posts are aligned, fill in the remaining dirt. Fill and tamp no more than 6 inches in the hole at one time to ensure good, solid embedment.

Select beams the size and length shown in the foundation plan of the working drawings. Moisture content should normally not exceed 19 percent. These beams are usually 2 by 10 or 2 by 12 inches in size. The lengths conform to the spacing of the posts. For example, posts spaced 8 feet apart will require 8- and 16-foot-long beams.

The outside beam can now be nailed in place. Starting along one side of the corner, even with the leveling string, nail one beam to the corner post and each crossing post. Initially, use only one twentypenny nail at the top of each beam, and don't drive it in fully. (The center should be left free to allow for a carriage bolt.)

Side beam ends should project beyond the post about 1½ inches or the thickness of the end header. When all outside beams and those along the center row of posts are erected, all final leveling adjustments can be made. In addition to the leveling cord from the layout stakes, use a carpenter's level and a straightedge to ensure that each beam is level, horizontal, and in line with the cord. Final nailing is now done on the first set of beams. Posts extending over the tops of the beams can be trimmed flush (Fig. 4-7).

The second set of beams on the opposite sides of the posts should not be installed. Because round posts vary in diameter, the facing on the second side has been delayed until this point. Use a strong cord or string and stretch it along the side of the foundation on the inside of the posts and parallel to the outside beams. This will establish the amount of notching and facing necessary for each post. Use a saw to provide a square notch to support the second beam.

All posts can be faced in this way, and the second beams nailed in place level with, and in the same way as, the outside beams. This facing is usually un-

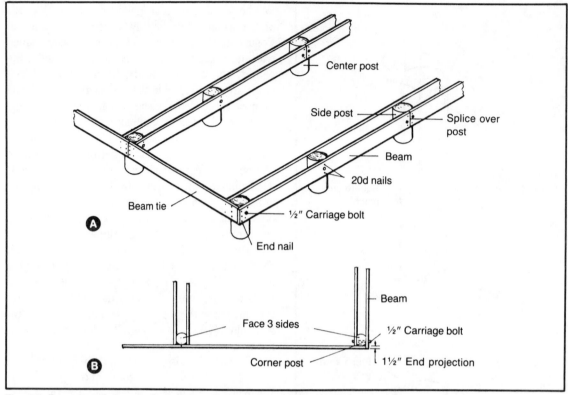

Fig. 4-7. Beam installation. A: Overall view. B: Plan view. Courtesy USDA

necessary when square posts are used. However, additional bolts are required when the beam does not bear on a notch. Joints of the headers are made over the center of the posts and staggered. For example, if an 8- and a 16-foot beam are used on the outside of the posts, stagger the joints by first using a 16-foot, then an 8-foot length, on the inside. Only one joint should be made at each support (Fig. 4-8).

Drill ½-inch holes through the double beams and posts at midheight of the beam. Install ½-by 8-, 10-, or 12-inch galvanized carriage bolts with the head on the outside and a large washer under the nut on the inside. When available at the correct moisture content, single nominal 4-inch-thick beams might be used to replace the 2-inch members (Fig. 4-9).

All poles and beams can be installed, and the final earth tamping done around the poles where required. A final raking and leveling is in order to ensure a good base for the soil cover if required. There

is now a solid level framework upon which to erect the floor joists.

Edge Piers

When masonry piers or wood posts are used along the edge of the building line instead of for overhang floor framing as previously described, the 8-inch poured footings are usually the same size as shown in Fig. 4-10(A). However, check the working drawings for the exact size. For masonry piers, the distance to the bottom of the footings is governed by the depth of frost penetration. This may vary from 4 feet in the northern states to less than 1 foot in southern areas. The wood posts normally require a 3-to-4-foot-deep hole. Align the masonry piers or posts so that the outside edges are flush with the outside of the building line (Fig. 4-10A and B).

Concrete block, brick, or other masonry, or poured concrete piers are constructed over the footings. Concrete block piers should be 8 by 16 inches

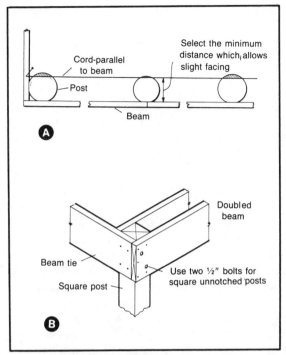

Fig. 4-8. Facing and fastening posts. A: Round posts. B: Square posts. Courtesy USDA

above the highest corner of the building area. Use any of the previously described leveling methods. Use a 22-gauge-by 2-inch-wide galvanized perforated or plain anchor strap for nailing into the beams. It should extend through at least two courses, filling the core when hollow masonry is used (Fig. 4-10A).

Use a prepared mortar mix with 3 or 3½ parts sand and ¼ part cement to each part of mortar or other approved mixes in laying up the masonry units. The wood post installation details are shown in Fig. 4-10. Nail anchor straps to each post and beam with twelvepenny galvanized nails.

Beams consisting of doubled 2-by-10- or 2-by-12-inch members (check the foundation plan of the working drawings) can now be assembled. Place them on the post or piers and make the splices at this location. Make only one splice at each pier. Nominal 2-inch members can be nailed together with tenpenny nails spaced 16 inches apart in two rows. Fabrication details at the corner and intersection with the center beam, and fastening of the beam tie (stringer) are shown in Fig. 4-10C.

Ledgers, used to support the floor joists, are nailed to the inside of the nailed beams. The sizes are 2 by 2, 2 by 3, or 2 by 4 inches as indicated in the foundation plan of the working drawings. Place ledgers so that the top of the joists will be flush with the top of the edge and center beams when bearing on the ledgers (Fig. 4-10A and B). Use sixteenpenny nails spaced 8 inches apart in a staggered row to fasten the ledger to the beam.

The foundation and beams are now in place ready for the assembly of the floor system.

An alternate method of providing footings for the treated wood foundation posts involves the use of temporary braces to position the posts while the concrete is poured (Fig. 4-11). After the holes are dug, posts of the proper length can be positioned and temporary braces nailed to them. Allow a minimum of 8 inches for footing depth. When posts are aligned and set to the proper elevation, pour concrete around them. After the concrete has set, holes can be filled (backfilled) and earth tamped firmly

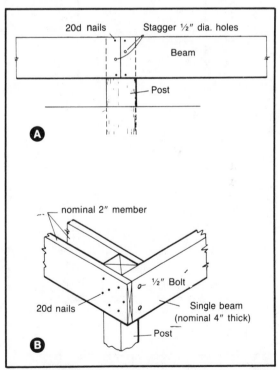

Fig. 4-9. Fastening beams to posts. A: Bolting. B: Single beam. Courtesy USDA

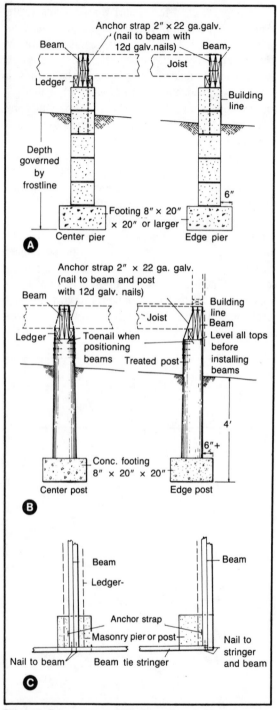

Fig. 4-10. Edge posts and masonry piers. A: Masonry edge piers. B: Edge post foundation. C: Corner and edge framing of beam. Courtesy USDA

around the posts. Construction of the floor framing can now begin.

This system of setting posts and pouring the concrete footings can also be used when a beam is located on each side of the posts (Fig. 4-7). Posts are placed in the holes, the beams nailed in their proper position, and the beams aligned and blocked to the correct elevation. After the concrete has set and the fill has been tamped in place, the beams can be bolted to the posts.

Termite Protection

In southern and central areas of the United States, and in the coastal states, termite protection must be considered in construction of crawlspace houses. In fact, it is common today to take this precaution regardless of geographic location. Pressure-treated lumber or poles are not affected by termites, but these insects can damage untreated wood.

An effective method of protection is soil poisons. Spraying soil with solutions of approved chemicals such as aldrin, chlordane, dieldrin, and helptachlor will provide protection for 10 or more years.

Many builders also use a termite protective metal that is nailed to the beams of the foundation.

A physical method of preventing entry of termites to untreated wood is by the use of termite shields (Fig. 4-12). These are made of galvanized iron, aluminum, copper, or other metal. They are located over continuous walls or over and around treated wood foundation posts. They are not effective if bent or punctured during or after construction.

Crawlspaces should have sufficient room so that an examination of poles and piers can be made each spring. These inspections provide safeguards against wood-destroying insects.

JOISTS FOR FLOOR SYSTEMS

The beams, or beams and ledger strips, and the foundations are now in place and joists can be installed. Size and lengths of the floor joists, as well as the species and spacing, are shown in the floor framing layout in the working plans for the house being built. The joists might vary from nominal 2 by 8 inches in size to 2 by 10 inches or larger where spans are long. Moisture content of floor joists and

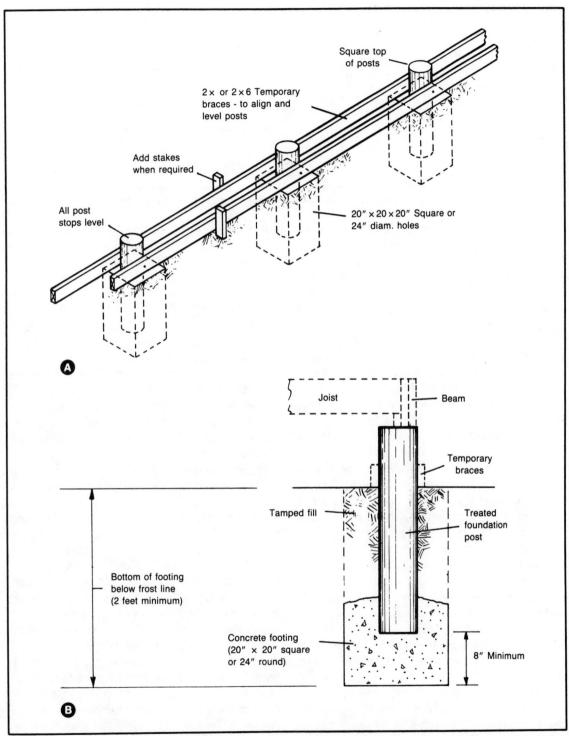

Fig. 4-11. Alternate method of setting edge foundation posts. A: Temporary bracing. B: Footing position. Courtesy USDA

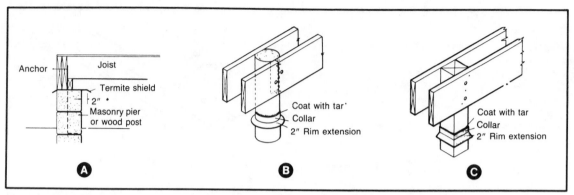

Fig. 4-12. Termite shields. A: On top of masonry or wood post. B: Round posts. C: Square posts. Courtesy USDA

other floor framing members should not exceed 19 percent when possible. Spacing of joists is normally 16 or 24 inches on center, so that 8-foot lengths of plywood for subfloor will span six or four joist spaces.

In low-cost houses, savings can be made by using plywood for subfloor, which also serves as a base for resilient tile or other covering. This can be done by specifying tongue-and-groove edges in a plywood grade of C-C plugged Exterior Douglas fir, southern pine, or similar species. Regular Interior Underlayment grade with exterior glue is also satisfactory. The matched edges provide a tight lengthwise joint, and end joints are made over the joists.

If tongue-and-groove plywood is not available, use square-edged plywood, and block between joists with 2 by 4s for edge nailing. Plywood subfloor also serves as a tie between joists over the center beam. Insulation should be used in the floor in some manner to provide comfort and reduce heat loss. It is generally used between or over the joists.

Single-floor systems can also include the use of nominal 1-by-4-inch matched finish flooring in species such as southern pine and Douglas fir and the lower grades of oak, birch, and maples in $^{25}/_{32}$-inch thicknesses. To prevent air and dust infiltration, cover joists with 15-pound asphalt felt or similar materials. The flooring is then applied over the floor joists and the floor insulation added when the house is enclosed.

When this single-floor system is used, some surface protection from weather and mechanical damage is required. A full-width sheet of heavy plastic or similar covering can be used, and the walls erected directly over the film. When most of the exterior and interior work is done, the covering can be removed and the floor sanded and finished.

Post Foundation With Side Overhang

The joists for a low-cost house are usually the third grade of such species as southern pine or Douglas fir, and are often 2 by 8 inches for spans of approximately 12 feet. If an overhang of about 12 inches is used for 12-foot lengths, the joist spacing normally can be 24 inches. Sizes, spacing, and other details are shown in the plans for each individual house.

The joists can be cut to length, using a butt joint over the center beam. For a 24-foot-wide house, cut each pair of joists to a 12-foot length, less the thickness of the end header joist—which is usually 1½ inches.

Position the edge of stringer joists on the beams with several other joists and nail the premarked headers to them with one sixteenpenny nail (or just enough to keep them in position). The frame, including the edge (stringer) joists and the header joists, is now the exact outline of the house. Square up this framework by using the equal diagonal method (Fig. 4-13). The overhang beyond the beams should be the same at each side of the house.

With eightpenny nails, toenail the joists to each beam they cross and the stringer joists to the beam beneath to hold the framework exactly square. Add the remaining joists and nail the headers into the ends with three sixteenpenny nails. Toenail the remaining joists to the headers with eightpenny nails.

When the center of a parallel partition wall is more than 4 inches from the center of the joists, add solid blocking between the joists. The blocking should be the same size as the joists and spaced not more than 3 feet apart. Toenail blocking to the joists with two tenpenny nails at each side.

In moderate climates, 1-inch blanket insulation should be sufficient to insulate the floor of crawl-space houses. It is usually placed between the joists in the same way that thicker insulation is installed. Another method consists of rolling 23-inch-wide, 1-inch-thick insulation across the joists, nailing or stapling it where necessary so it is stretched with tight edge joints. Insulation of this type should have strong, damage-resistant covers.

Use tenpenny ring-shank nails to fasten the plywood to the joists rather than the eightpenny nails that are normally used. This will minimize nail movement or "nail pops" that could occur during mois-

ture changes. The vapor barrier of the insulation should be on the upper side toward the subfloor. Place 2-inch and thicker blanket or batt insulation between the joists any time after the floor is in place, preferably when the house is near completion.

When the house is 20, 24, 28, or 32 feet wide, the first row of tongue-and-groove plywood sheets should be 24 inches wide, so that the butt joints of the joists at the center beam are reinforced with a full 48-inch-wide piece (Fig. 4-14). This plywood is usually ⅝ to ¾ inch thick when it serves both as subfloor and underlayment.

Rip 4-foot-wide pieces in half and save the other halves for the opposite side. Place the square, sawed edges flush with the header, and nail the plywood to each crossing joist and header with eightpenny common nails spaced 6 to 7 inches apart at edges and at intermediate joists (Fig. 4-13). Joints in the next full 4-foot widths of plywood should be broken

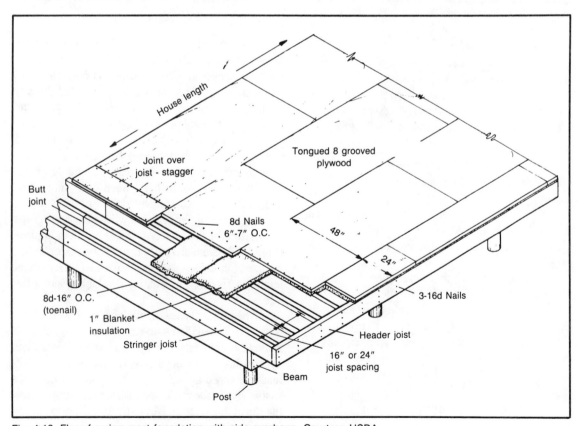

Fig. 4-13. Floor framing, post foundation with side overhang. Courtesy USDA

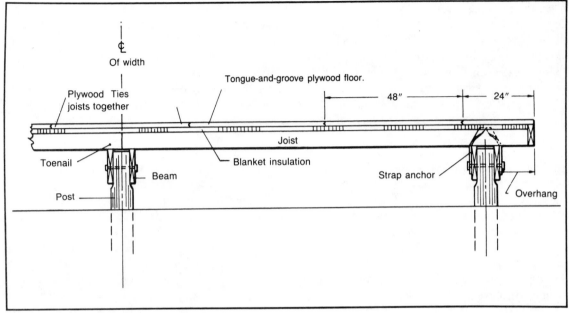

Fig. 4-14. Floor framing details, post foundation. Courtesy USDA

by starting at one end with a 4-foot-long piece. End joints will thus be staggered 48 inches. End joints should always be staggered at least one joist space, 16 or 24 inches. Be sure to draw up the tongue-and-groove edges tightly. Use a chalked snap-string to mark the position of the joists for nailing.

Edge Foundation

When edge piers or wood posts are used with edge support beams, the joists can be cut to length to fit snugly between the center and outside beams. Then they will rest on the ledger strips. Toenail each end of the beams with two eightpenny nails on each side (Fig. 4-15).

When applying the plywood subfloor to the floor framing, start with full 4-foot-wide sheets rather than 2-foot-wide pieces used for the side overhang framing (Fig. 4-16). The nail-laminated center beam provides sufficient reinforcing between the ends of joists. Apply the insulation and nail the plywood the same way as outlined in the previous section.

Insulation Between Joists

Thicker floor insulation than the 1-inch blanket is re-

quired for most houses. This will be indicated in the floor framing details for the working drawings or in the specifications. This insulation is normally used between the joists. The subfloor is nailed directly to the joists and the insulation placed between the joists after the subfloor is in place.

Use friction-fit (or similar insulating batts) 15 inches wide for joists spaced 15 inches on center. Use 28-inch-wide batts for joists spaced 24 inches on center. Friction-fit batts need little support to keep them in place. Small "dabs" of asphalt roof cement on the upper surfaces when installing against the bottom of the plywood will keep them in place (Fig. 4-17). Standard batts can also be placed in this manner, but somewhat closer spacing of the cement might be required in addition to stapling along the edges.

When other types of subfloor, such as diagonal boards, are specified some kind of overlay or finish is usually required. If tongue-and-groove flooring is applied directly to and across the joists, a tie is normally required at the center batt joints of the floor joists. This is accomplished with a metal strap across the top of the joists or 1-by-4-by-20-inch wood strips (scabs) nailed across the faces of each set of joists

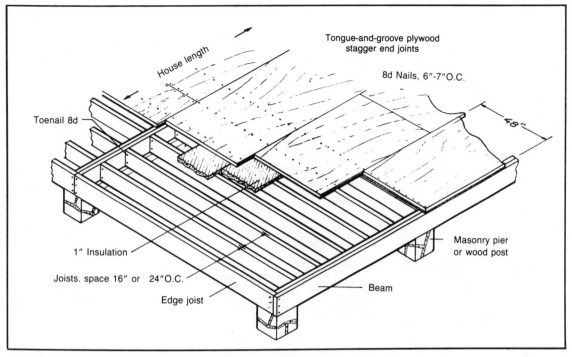

Fig. 4-15. Floor framing for edge. Courtesy USDA

at the joint with eightpenny nails. When plywood subfloor is used, the sheets are centered over the center beams and joist ends to provide this tie for overhang floor framing.

Finally, if the plywood is likely to be exposed for any length of time before enclosing the house, use a brush coat (or squeegee application) of water-repellent preservative. This will not only repel moisture but will prevent or minimize any surface degradation.

FRAMED WALL SYSTEMS

Exterior sidewalls, and in some designs an interior wall, normally support most of the roofloads and also serve as a framework for attaching interior and exterior coverings. When roof trusses spanning the entire width of the house are used, the exterior sidewalls carry both the roof and ceiling loads. Interior partitions then serve mainly as room dividers. When ceiling joists are used, interior partitions usually sustain some of the ceiling loads.

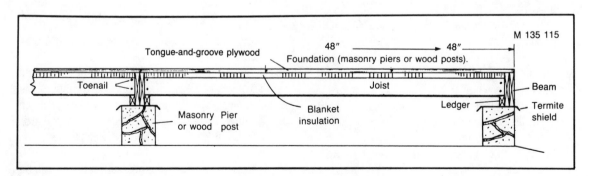

Fig. 4-16. Floor framing details, edge pier foundation. Courtesy USDA

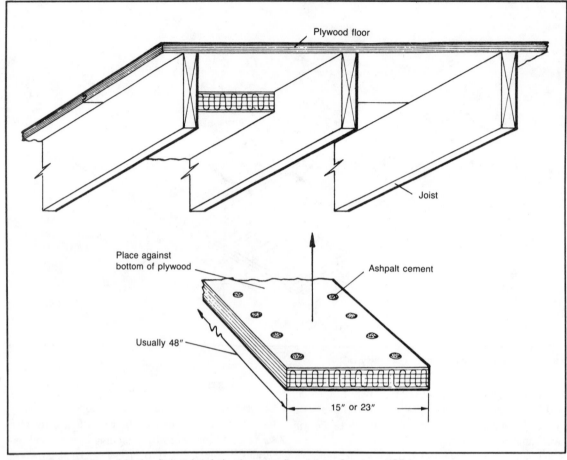

Place against
bottom of plywood

Ashpalt cement

Usually 48"

15" or 23"

Fig. 4-17. Application of friction-fit batt insulation between floor joists. Courtesy USDA

The exterior walls of a wood-frame house normally consist of studs, interior and exterior coverings, windows and doors, and insulation. Moisture content of framing members usually should not exceed 19 percent.

The framework for a conventional wall consists of nominal 2-by-4-inch members used as top and bottom plates, as studs, and as partial (cripple) studs around openings. Studs are generally cut to lengths for 8-foot walls when subfloor and finish floors are used. This length depends on the thickness and number of wall plates (normally single bottom and double top plates). Studs can often be obtained from lumber dealers in a precut length. Two 2-by-6-inch members are used for spans up to 4½ feet and two 2-by-8-inch members for openings from 4½ to about

6½ feet. Headers are normally cut 3 inches (two 1½-inch stud thicknesses) longer than the rough opening width unless the edge of the opening is near a regular-spaced stud.

Framing of Sidewalls

The exterior framed walls, when erected, should be flush with the outer edges of the plywood subfloor and floor framing. Thus, the floor can be used both as a layout area and for horizontal assembly of the wall framing. When completed, the entire wall can be raised in place in "tilt-up" fashion, the plates nailed to the floor system, and the wall plumbed and braced.

The two exterior sidewalls of the house can be framed first and the exterior end walls framed later.

Cut two sets of plates for the entire length of the house, using 8-, 12-, or 16-foot lengths, staggering the joints. Make joints at the centerline of a stud (on center). Starting at one end, mark each 16 or 24 inches, depending on the spacing of the studs. Also mark the centerlines for windows, doors, and partitions. These measurements are given on the working drawings. These are centerline (on center) markings, except for the ends. With a small square, mark the location of each stud with a line about ¾ inch on each side of the centerline mark (Fig. 4-18).

Studs can now be cut to the correct length. When using a slow-slope roof with wood decking (which also serves as a ceiling finish), the stud length for an 8-foot wall height with single plywood flooring should be 95½ inches, less the thickness of three plates. For plate 1½ inches thick, this will mean a stud length of 91 inches. When ceiling joists or trusses are used with the single plywood floor, this length can be about 92⅛ inches. These measurements are primarily to provide for the vertical use of 8-foot lengths of drywall sheet materials for the walls with the ceiling flush in place. Cornerposts can be made up beforehand by nailing two short 2-by-4-inch blocks between two studs. Use two twelvepenny nails at each side of each block (Fig. 4-19).

Begin fabrication of the wall by fastening the bot-

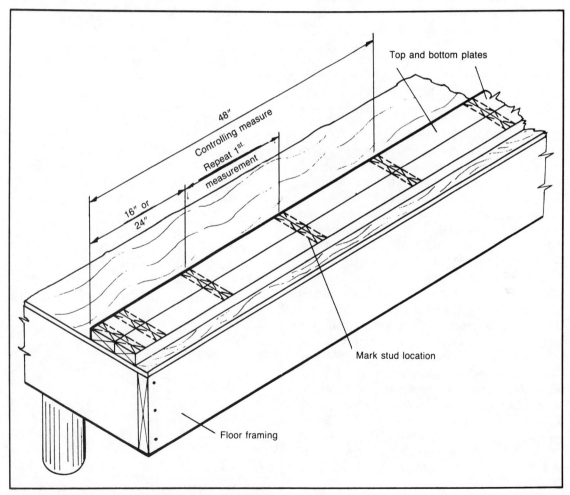

Fig. 4-18. Marking top and bottom plate. Courtesy USDA

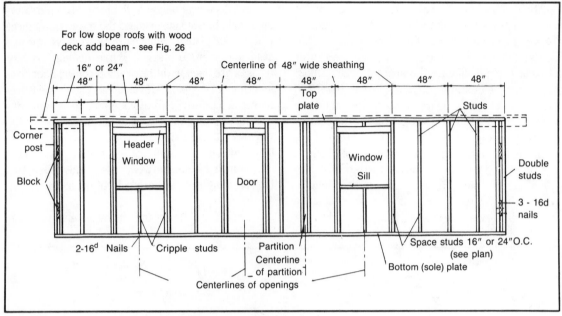

For low slope roofs with wood deck add beam - see Fig. 26

16" or 24"

48" 48" 48" 48" 48" 48" 48" 48"

Centerline of 48" wide sheathing

Top plate

Studs

Corner post

Header

Window

Block

Window

Sill

Double studs

3 - 16d nails

Door

2-16ᵈ Nails

Cripple studs

Partition Centerline of partition

Centerlines of openings

Bottom (sole) plate

Space studs 16" or 24"O.C. (see plan)

Fig. 4-19. Framing layout of typical wall. Courtesy USDA

tom plate and the first top plate to the ends of each cornerpost and stud, using two sixteenpenny nails in each member. As studs are nailed in place, make provisions for framing the openings for windows and doors. Studs should be located to form the rough openings, the sizes of which vary with the types of windows selected. The rough openings are framed by studs which support the window and floor headers or lintels. A full-length stud should be located at each side of these framing studs (Fig. 4-20).

The following allowances are usually made for rough opening widths and heights for doors and windows. Half of the given width should be marked on each side of the centerline of the opening, which has previously been marked on top and bottom plates.

Double-hung window (single unit): Rough opening width = glass width plus 6 inches. Rough opening height = total glass height plus 10 inches.

Casement window (two sash): Rough opening width = total glass width plus 11¼ inches. Rough opening height = total glass height plus 6⅜ inches.

Exterior doors: Rough opening width = width of door plus 2½ inches. Rough opening height = height of door plus 3 inches.

Clearances, or rough-opening sizes, for typical

double-hung windows, for example, are shown in Table 4-1.

Figure 4-20 shows the height of the window and door headers above the subfloor when doors are 6 feet, 8 inches high and finish floor is used. When only resilient tile is used over flooring made up of a sin-

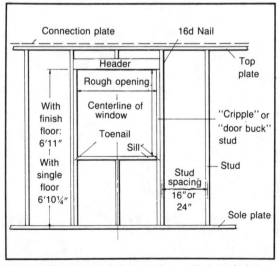

Connection plate

16d Nail

Header

Top plate

Rough opening

With finish floor: 6'11"

Centerline of window

"Cripple" or "door buck" stud

Toenail

Sill

With single floor 6'10¼"

Stud

Stud spacing 16"or 24"

Sole plate

Fig. 4-20. Framing at window opening and height of window and door headers. Courtesy USDA

Table 4-1. Frame Opening Sizes for Double-Hung Windows.

Window glass size (each sash)		Rough frame opening size	
Width Inches	Height Inches	Width Inches	Height Inches
24	16	30	42
28	20	34	50
32	24	38	58
36	24	42	58

Courtesy USDA

gle layer of material, the framing height for windows and doors should be 6 feet 10¼ inches for 6-foot, 8-inch doors. The sizes of headers should be the same as those previously outlined.

Doubled headers can be fastened in place with two sixteenpenny nails through the stud into each member. Nail cripple (door buck) studs supporting the header on each side of the opening to the full stud, with twelvepenny nails spaced about 16 inches apart and staggered. Toenail the sill and other short (cripple) studs in place with two eightpenny nails at each side when endnailing is not possible.

Double studs should normally be used on exterior walls where intersecting interior partitions are located. This can be accomplished with spaced studs (Fig. 4-21) and provides nailing surfaces for interior covering materials. Blocking with 2-by-4-inch members placed flatwise between studs spaced 4 to 6 inches apart in the exterior wall might also be used to fasten the first partition stud. Space blocks about 32 inches apart. When a low-slope roof with gable overhang is used with wood decking, a beam extension is required at the top plates (Figs. 4-19 and 4-22).

Erecting Sidewalls

When the sidewalls are completed, they can be raised in place. Nail several short 1-by-6-inch pieces to the outside of the beam to prevent the wall from sliding past the edge. Fasten the bottom plate to the floor framing with sixteenpenny nails spaced 16 inches apart and staggered when practical. The wall can now be plumbed and temporary bracing added to hold it in place in a true vertical position.

Bracing may consist of 1-by-6-inch members nailed to one face of a stud and to a 2-by-4-inch block that has been nailed to the subfloor. Braces should be at about a 45-degree angle. If the wall framing is squared and braced, the panel siding or exterior covering can be fastened to the studs while the walls are still on the subfloor. In addition, window frames can be installed before erection of the wall. These processes are covered in following sections.

End Walls: Moderate-Slope Roof

The exterior walls for a gable-roofed house may be assembled on the floor in the same general manner as the sidewalls, with a bottom plate and single top plate. However, the total length of the wall should be the exact distance between the inside of the exterior sidewalls already erected. Furthermore, only one end stud is used rather than the doubled cornerposts (Fig. 4-23).

Frame window and door openings as outlined for the exterior sidewalls. When 48-inch-wide panel siding is used for the exterior—serving both as sheathing and siding, for example—the stud spacing should conform to the type of covering used. The center of the second stud in this wall should be 16 or 24 inches from the outside of the panel-siding material. Use this method of spacing from each corner toward the center. Any adjustments required because of sheet-material width should be made at a center window or door (Fig. 4-24).

End walls are erected in the same manner as the sidewalls, with the bottom plate fastened to the floor framing. These walls also must be plumbed and braced. Nail the end studs at each side to the cornerposts with sixteenpenny nails spaced 16 inches apart. The upper top plate can now be added. It extends across the sidewall plate.

The framing for the gable-end portion of the wall is often done separately. Studs may be toenailed to the upper top wall plate, or an extra 2-by-6-inch bottom plate can be used, which provides a nailing surface for ceiling material in the rooms below. The top members of the gable wall are not plates; they are rafters that form the slope of the roof. Studs are notched to fit or may be used flatwise. Use the roof slope specified in the working drawings.

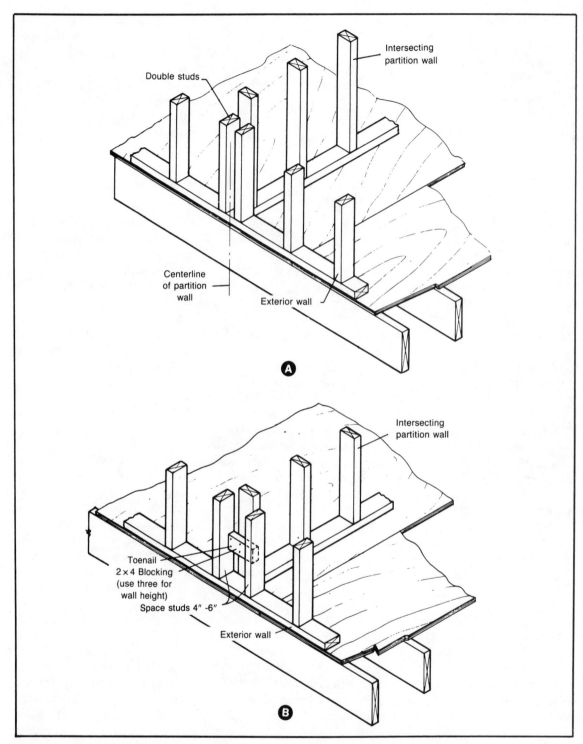

Fig. 4-21. Intersecting walls. A: Double studs. B: Blocking between studs. Courtesy USDA

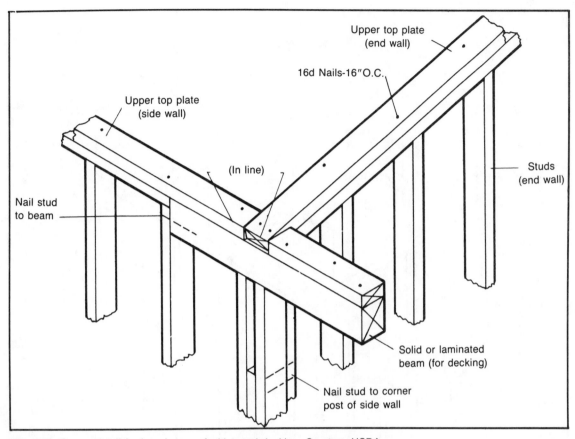

Fig. 4-22. Corner detail for low-slope roof with wood decking. Courtesy USDA

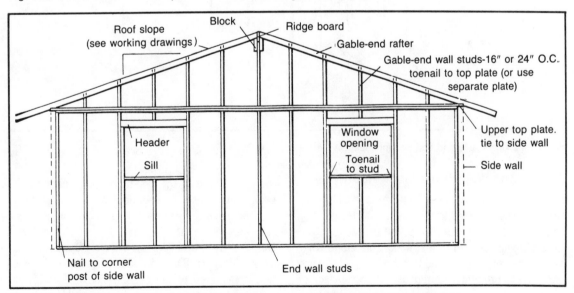

Fig. 4-23. End wall framing for regular slope roof (for trusses or rafter-type). Courtesy USDA

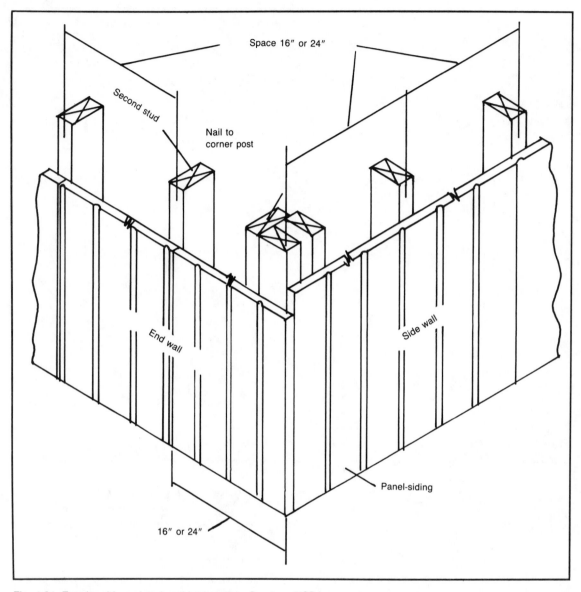

Space 16″ or 24″

Second stud

Nail to
corner post

End wall

Side wall

Panel-siding

16″ or 24″

Fig. 4-24. Exterior side and end wall intersection. Courtesy USDA

End Walls: Low-Slope Roof

End walls for a low-slope roof are normally constructed with balloon framing. In this design, the
studs are full-length from the bottom plate to the
top (rafter) plates which follow the roof slope (Fig.
4-25). The stud spacing and framing for windows are
the same as previously outlined. The top surface of
the upper plate of the end wall should be in line with

the outer edge of the upper top plate of the sidewall. Thus, when the roof decking is applied, bearing and nailing surfaces are provided at end and
sidewalls. A beam extension beyond the end wall
supports the wood decking when a gable-end overhang is desired. This can be a 4-by-6-inch member
that is fastened to the second sidewall stud.

The lower top plate of the end wall is nailed to

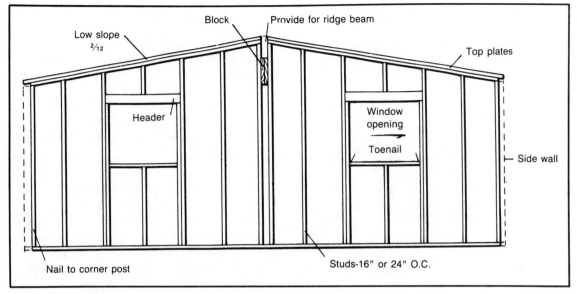

Fig. 4-25. Framing for end wall (low-slope roof). Courtesy USDA

the end of the studs before the upper top plate is fastened in place. For attaching the upper plate, use sixteenpenny nails spaced 16 inches apart and staggered. Use two nails over the cornerposts of the sidewalls.

To provide for a center ridge beam which supports the wood decking inside the house, the area should be framed (Fig. 4-26). After the beam is in place, use twelvepenny nails through the stud on each side of the beam. The size of the ridge beam

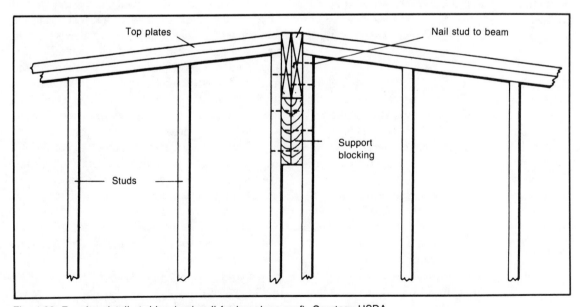

Fig. 4-26. Framing detail at ridge (end wall for low-slope roof). Courtesy USDA

is shown on the working plans for each house which is constructed using this method. When decking is used for a gable overhang, the beam extends beyond the end walls.

Interior Walls

Interior walls in conventional construction (with ceiling joists and rafters) are erected in the same manner and at the same height as the outside walls. In general, assembly of interior stud walls is the same as outlined for exterior walls. Locate the center load-bearing partition so that ceiling joists require little, or no, wasteful cutting. Cross partitions are usually not load-bearing and can be spaced as required for room sizes.

Space studs according to the type of interior covering material to be used. When studs are spaced 24 inches on center, the thickness of gypsum board, for example, must be ½ inch or greater. For 16-inch stud spacing, a thickness of ⅜ inch or greater can be used. Details of a typical intersection of interior walls are shown in Fig. 4-27. Load-bearing partitions should be constructed with nominal 2-by-4-inch studs, but 2-by-3-inch studs may also be used for nonload-bearing walls. Doorway openings can also be framed with a single member on each side in nonload-bearing walls. Single top plates are commonly used on nonload-bearing interior partitions.

ROOF SYSTEMS

Roof trusses require no load-bearing interior partitions, so location of the walls and size and spacing of studs are determined by the house design and by the type of interior finish. The bottom chords of the trusses are often used to tie in with crossing partitions where required.

When a low-slope roof is used with wood decking, a full-height wall or a ridge beam is required for support at the center (Fig. 4-28). The ridge beam may span from an interior center partition to an outside wall, forming a clear open area beneath. Cross or intersecting walls are full height with a sloping top plate. In such designs, follow this sequence: Erect exterior walls, center wall, and ridge beam.

Apply roof decking. Install other partition walls. Size and spacing of the studs in the cross walls are usually based on the thickness of the covering material, because no roof load is imposed on them. These spacings and sizes are part of the working drawings.

The upper top plates (connecting plates) are used to tie the wall framing together at corners, at intersections, and at crossing walls. The upper plate crosses and is nailed to the plate below (Figs. 4-22 and 4-27). Use two sixteenpenny nails at each intersection. Nail the remainder of the upper top plate to the lower top plate with sixteenpenny nails spaced 16 inches apart in a staggered pattern.

The primary function of a roof is to provide protection to the house in all types of weather with a minimum of maintenance. A second consideration is appearance; a roof should add to the attractiveness of the home. Happily, a roof with a wide overhang at the cornice and the gable ends not only enhances appearance, but provides protection to side and end walls. Therefore, even in lower cost houses—when the style and design permit—wide overhangs are desirable. Though they add slightly to the initial cost, savings in future maintenance usually merit such an extension.

As described, the two types of roofs commonly used for houses are the low-slope and the pitched roof. The flat or low-slope roof combines ceiling and roof elements as one system, which allows them to serve as interior finish, as a fastening surface for finish, and as an outer surface for application of the roofing. The structural elements are arranged in several ways by the use of ceiling beams or thick roof decking, which spans from the exterior walls to a ridge beam or center bearing partition. Roof slope is usually designated as some ratio of 12. For example, a 4-in-12 roof slope has a 4-foot vertical rise for each 12 feet of horizontal distance.

The pitched roof, usually in slopes of 3 in 12 and greater, has structural elements in the form of rafters and joists or trusses (trussed rafters). Both systems require some type of interior ceiling finish, as well as roof sheathing. With slopes of 8 in 12 and greater, it is possible to include several bedrooms on the second floor when provisions are made for floor loads, a stairway, and windows.

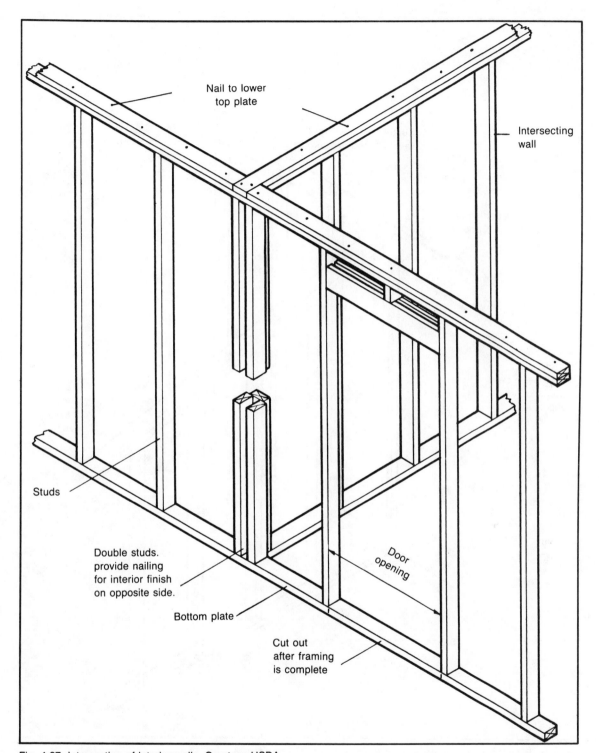

Nail to lower
top plate

Intersecting
wall

Studs

Double studs.
provide nailing
for interior finish
on opposite side.

Bottom plate

Door
opening

Cut out
after framing
is complete

Fig. 4-27. Intersection of interior walls. Courtesy USDA

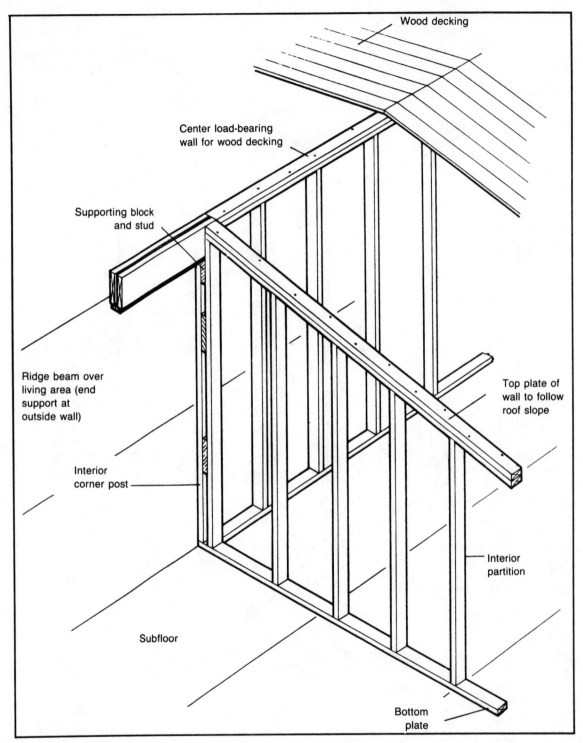

Fig. 4-28. Ridge beam and center wall for low-slope roof. Courtesy USDA

Wood decking

Center load-bearing
wall for wood decking

Supporting block
and stud

Ridge beam over
living area (end
support at
outside wall)

Interior
corner post

Subfloor

Top plate of
wall to follow
roof slope

Interior
partition

Bottom
plate

Low Slope Ceiling Beam Roof

One of the framing systems for a low-slope roof consists of spaced rafters (beams or girders) that span from the exterior sidewalls to a ridge beam or a center load-bearing wall. The rafter can be doubled, spaced 4 feet apart, and exposed in the room below, providing a pleasing beamed ceiling effect. Dressed and matched V-groove boards can be used for roof sheathing and be exposed to the room below. When plywood or other unfinished sheathing is used, a ceiling tile or other prefinished wallboard can be fastened to the under surface. Such materials also serve as insulation. Thus, a very attractive ceiling can be provided using a light color for the ceiling and a contrasting stain on the beams. This type of framing can be varied by spacing single rafters on 16- or 24-inch centers. Separate covering materials would normally be used for the roof sheathing and for the ceiling, with flexible insulation between.

The size and spacing details for ceiling beams are shown on the working drawings for each house design. For example, when beams are doubled, spaced 48 inches apart, and the distance from outer wall to interior wall is about 11½ feet, two 2-by-8-inch members are satisfactory for most of the construction species such as Douglas fir, southern pine, and hemlock. Use of some wood species, such as the soft pines, will require two 2-by-10-inch members for 48-inch spacing. When a spacing of 32 inches is desirable for appearance, two 2-by-6-inch members of the second grade of Douglas fir or southern pine are satisfactory. In some of the species, such as southern pine and Douglas fir, a solid 4-by-6-inch member provides sufficient strength for a 48-inch spacing over an 11½-foot span.

Construction

Let us assume that the details in the working plans specify doubled 2-by-8-inch ceiling beams spaced 48 inches on center. When the beams do not extend beyond the wall, a look-out member is required for the cornice overhang. The center wall or ridge beam is in place, so the roof slope, which might vary between 1½ in 12 to 2½ in 12, has been established. As previously outlined in the section on wall systems, the ceiling beams should normally be erected before cross walls are established. Thus, the exterior sidewalls and the load-bearing center wall, all well braced and plumbed, are all that is required to erect the ceiling beams.

One way to support the beams at the center bearing wall or ridge beam is to fasten a 2-by-3-inch block to the stud wall. Another is to use a metal joist hanger. A third is to notch the beam ends and fasten them to the stud. The first two may be used for either a stud wall or a ridge beam. Fastening at the outside wall is generally the same for all three methods.

Cut the first or sample beam to serve as a pattern in cutting the remainder of the members. Figure 4-29A shows the location of the ceiling beam with respect to the exterior and loadbearing center walls. When beams themselves do not serve as roof extensions, they can be assembled by nailing a 2-by-6-inch lookout (roof extension) at the outer wall and a 2-by-4-inch spacer block at the interior center wall (Fig. 4-29B). Use two twelvepenny nails on each side of the block and twelvepenny nails spaced 6 inches apart for the 2-by-6-inch lookout member. When using a block nailed to the stud to support the interior beam end, notch the ends. (Fig. 4-29C)

Details at Center Wall or Beam

The first system of connecting the inside end of the ceiling beam is most adaptable to the ridge-beam construction. It consists of fastening a nominal 2-by-3-inch block to the beam with 4½-inch lag screws (Fig. 4-30A). The 2-by-3 should be the same depth as the ceiling beam. The beam ends are then bolted to the 2-by-3-inch block with ⅜-by-5-inch carriage bolts.

Using joist hangers (Fig. 4-30B) to fasten the inside end of the ceiling beam is another method most adaptable to a ridge-beam system. Hangers are fastened to the ridge-beam face, the ceiling beams dropped in place, and the hangers nailed to the beams. Eight or tenpenny nails are commonly used for nailing. Hangers will be exposed, but can be painted to match the color of the beams.

A third method that can be used at a center bearing wall involves notching the ends of the ceil-

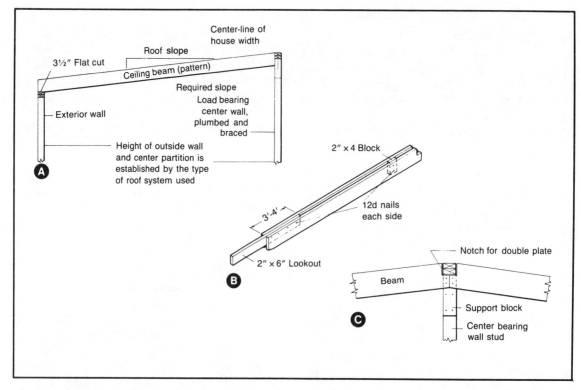

Fig. 4-29. Ceiling beam location and support methods. A: Layout of typical beam. B: Assembly of beam with roof extension (lookout). C: Notch for stud support at center bearing wall. Courtesy USDA

ing beams (Fig. 4-30C). When the beam is in place, it can be face-nailed to the stud at each side with two twelvepenny nails. short 2-by-4-inch support blocks are then nailed at each side of the stud with twelvepenny nails.

When solid 4-by-6-inch or larger members are used as ceiling beams in place of the doubled members, the joist hanger is probably the most suitable method of supporting the inside ends of the beams. If the solid or laminated beams are to be stained, take care to prevent hammer marks.

Nailing Ceiling Beams at Exterior Walls

The ceiling beams are normally fastened to the top plate of the outside walls with nails. In windy areas, some type of strapping or metal bracket might be necessary. Toenail the ceiling beams to the top of the outside wall with two eightpenny nails at each side and a tenpenny nail at the ends (Fig. 4-31). To

provide nailing for panel siding and interior finish, fasten 2-by-4-inch nailing blocks between the ceiling beams. Toenail with eightpenny nails at each edge and face.

Roof Sheathing

Ceiling beams and roof extensions are now in place and ready for installation of the roof sheathing. Roof sheathing can consist of 1-by-6-inch tongue-and-groove (V-grooved) lumber with $^{25}/_{32}$-inch fiberboard nailed over the top for insulation. Use two eightpenny nails for each board at each ceiling beam. The insulation fiberboard can be nailed in place with $1\frac{1}{2}$-inch roofing nails spaced 10 inches apart in rows 24 inches on center. Cross-sections of the completed wall and roof framing are shown in Fig. 4-32. A nominal 1-inch member about $3\frac{3}{4}$ inches wide may be used to case the undersides of the beams.

A gable-end extension of 16 inches or less can

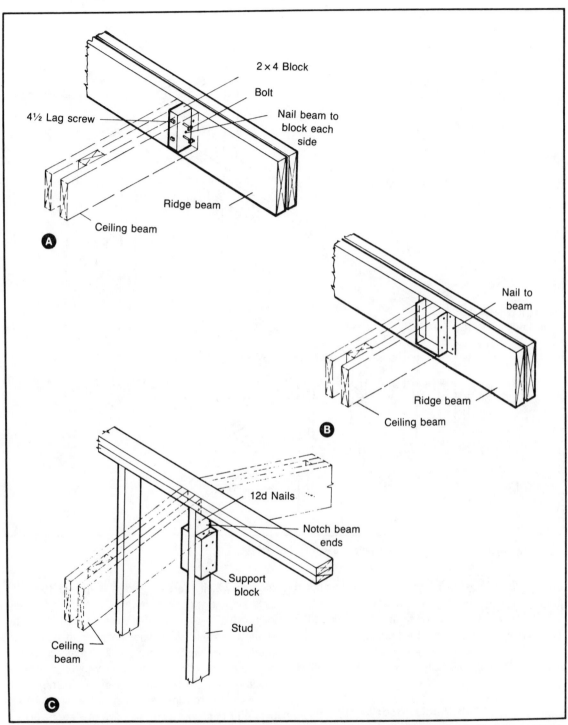

Fig. 4-30. Beam connection to ridge-beam or load-bearing wall. A: Block support. B: Joist hanger. C: Stud support block. Courtesy USDA

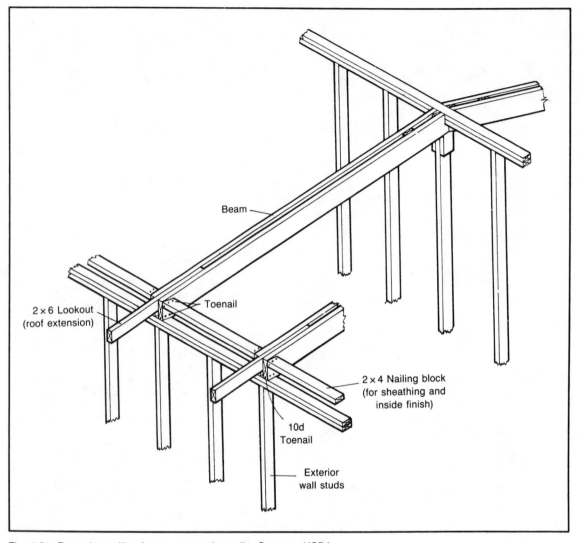

Fig. 4-31. Fastening ceiling beams at exterior walls. Courtesy USDA

be supported by extending the dressed and matched V-edge roof boards (Fig. 4-33). A 2-by-2 or larger member (fly rafter) is nailed to the underside of the boards and serves to fasten the facia board and molding. The V-groove of the underside of the 1-by-6-inch roof sheathing serves as a decorative surface.

Rafter-Joist Roof

Another type of construction for low-slope roofs similar to the ceiling beam framing is the rafter-joist roof, in which the members are spaced 16 or 24 inches apart and serve both as rafters and ceiling joists. (Fig. 4-34). Members may be 2 by 8 or 2 by 10 inches in size. The space between joists is insulated, allowing space for a ventilating airway. Gypsum board or other types of interior finish can be nailed directly to the bottoms of the joists.

Rafter extensions serve as nailing surfaces for the soffit of a closed cornice. When a open cornice is used, a nailing block is required over the wallplates and between rafters for the siding or frieze board.

The inside ends of the rafter-joists bear on an interior load-bearing wall. Beams are toenailed to the plate with eightpenny nails on each side. A 1-by-4-inch wood or ⅜-inch plywood scab is used to connect opposite rafter-joists. This fastens the joist together and serves as a positive tie between the exterior sidewalls.

Low-Slope Wood-Deck Roof

A simple method of covering low-slope roofs is with wood decking. Decking should be strong enough to span from the interior center wall or beam to the exterior wall. Decking can also extend beyond the wall to form an overhang at the eave line (Fig. 4-35A). This system requires dressed and matched

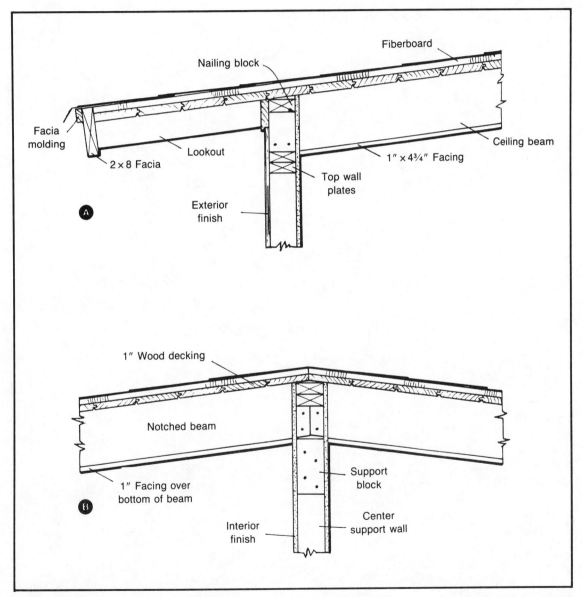

Fig. 4-32. Cross sections of completed walls and roof framing. A: Section through exterior wall. B: Section through center wall. Courtesy USDA

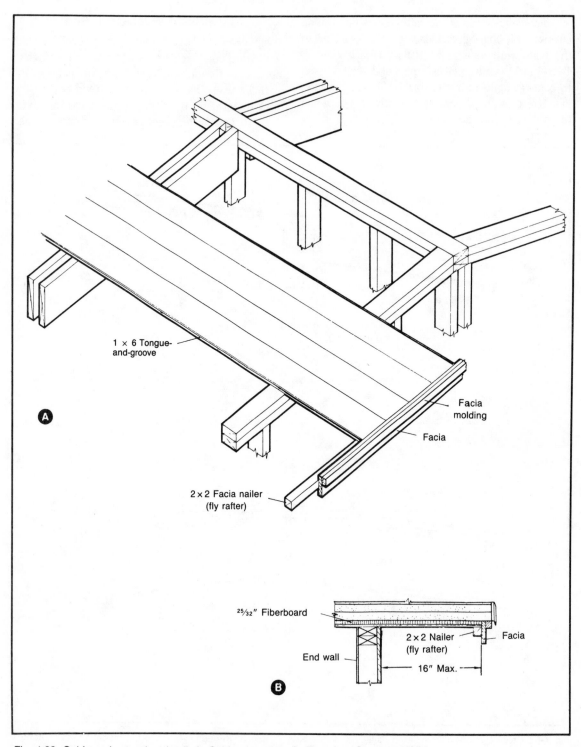

1 × 6 Tongue-
and-groove

Facia
molding

Facia

2 × 2 Facia nailer
(fly rafter)

A

²⁵⁄₃₂″ Fiberboard

2 × 2 Nailer
(fly rafter)

Facia

End wall

16″ Max.

B

Fig. 4-33. Gable-end extension detail. A: Gable extension. B: Fly rafter. Courtesy USDA

nominal 2-by-6-inch solid or laminated decking (cedar or similar species) for spans of about 12 feet. While this system requires more material than the beam and sheathing system, the labor involved at the building site is usually much less.

When gable-end extension is desired, some type of support is required beyond the end wall line at plate and ridge. This is usually accomplished by the projection of a small beam at the top plate of each sidewall (Figs. 4-19 and 4-22), and at the ridge. Often the extension of the double top plate of the sidewall is sufficient. Depending on the type and thickness, the decking must sometimes be in one full-length piece without joints unless there are in-termediate supports in the form of an interior partition. When such an interior wall is present, a butt joint can be made over its center.

Figure 4-35 illustrates the method of applying wood decking. This type of wood decking usually has decorative V-edge face, which should be placed down. A light-colored stain or other finish can be applied to the wood decking before it is installed. Prefinished members can also be obtained. Face-nail each 2-by-6-inch decking member to the ridge beam or center wall and to the top plates of the exterior wall with two sixteenpenny ring-shank nails. In addition, toenail sixpenny finish nails along each joint on 2- or 3-foot centers (Fig. 4-35B). A 40-degree

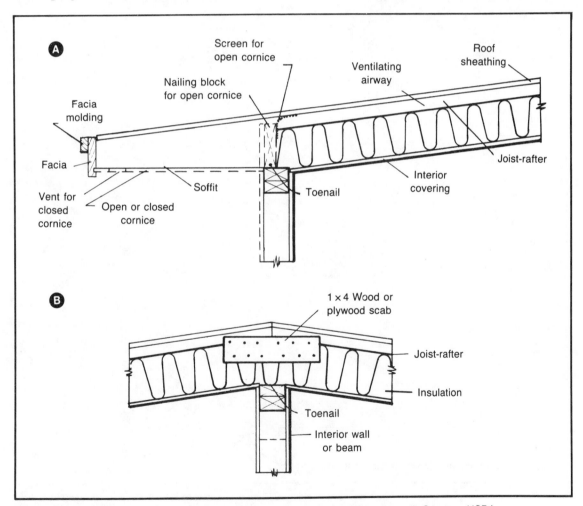

Fig. 4-34. Rafter-joist construction. A: Detail at exterior wall. B: Detail at interior wall. Courtesy USDA

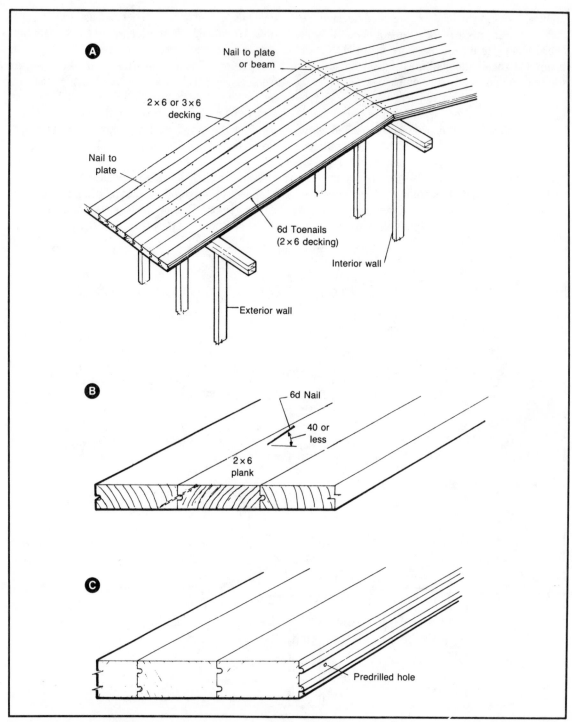

Fig. 4-35. Wood-deck construction. A: Installing wood decking. B: Toenailing horizontal joint. C: Edge nailing 3-by-6-inch solid decking. Courtesy USDA

angle or less should be used so that the nail point does not penetrate the underside. Drive nailheads flush with the surface.

When nominal 3-by-6-inch decking in solid or laminated form is required, it is face-nailed with two twentypenny ring-shank nails at center and outside wall supports. Solid 3-by-6-inch decking usually has a double tongue-and-groove and is provided with holes between the tongues for horizontal edge nailing (Fig. 4-35C). This edge nailing is done with 7- or 8-inch-long ring-shank nails, often furnished by manufacturers of the decking. Laminated decking can be nailed along the lengthwise joints with sixpenny nails through the groove and tongue. Space nails about 24 inches apart.

Decking support at the load-bearing center wall and at the ends of the sidewalls may also be provided by an extension of the top plates. (Fig. 4-36). The sides of the members can be faced later, if desired, with the same material used for siding or with 1-by-6-inch and 1-by-8-inch members.

Insulation for Low-Slope Roofs

It is considerably more difficult and costly to provide the low-slope roof with adequate insulation, except when the rafters-joist system is used with both interior and exterior covering. Even then the joists must have adequate depth to provide space for in-

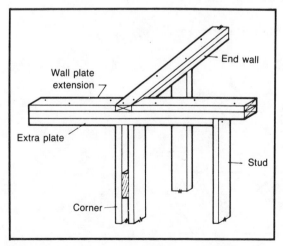

Fig. 4-36. Extension of wallplate for decking support. Courtesy USDA

sulation. Therefore, low-slope roofs are most economical in extremely mild climates or in houses that are used seasonally, where heating and air conditioning requirements are minimal.

Some insulation can be added economically by using $25/32$-inch insulating board sheathing placed over the wood decking. Expanded foam in sheet form, used as a base for the $1/2$-inch insulation board or tile on the underside of the decking, will provide increased resistance to heat loss. Rigid foam insulation can also be used on top of the decking, but $3/8$-inch plywood is required over the foam to provide a nailing base for the shingles.

The U-value of a building component is a measure of resistance to heat loss or gain. The lower the number, the more effective the insulation. Figure 4-37 illustrates the value of adding various insulating materials to the basic wood decking.

The method of installing the $25/32$-inch insulating sheathing over the wood decking is relatively simple. Lay the 4-by-8-foot sheets horizontally across the decking. Use $1\frac{1}{4}$- or $1\frac{1}{2}$-inch roofing nails spaced 10 inches apart in rows 24 inches apart along the length.

Insulating board or tile in $1/2$-inch thickness and the expanded foam insulation are installed with wallboard adhesive designed for these materials. Manufacturers normally recommend the type and method of application. with $1/2$-by-12-by-12-inch tile, for example, use a small amount of adhesive in each corner and apply hand pressure as the tile is placed. When tile is tongue-and-groove, stapling is the usual method of installation. Larger sheets of $1/2$-inch insulating board might require a combination of glue and nailing. Expanded foam insulation is ordinarily installed with approved adhesives.

When 1-inch wood decking is used over the joist-beam system, $25/32$-inch insulating board or tile on the inside is normally recommended. A 1-inch thickness of expanded foam insulation under the $1/2$-inch tile provides even better insulation. When the inner face of the decking is to be covered, lower grade 2-by-6-inch decking is commonly used.

Rigid foam over the deck is usually applied with roofing nails that penetrate the deck a minimum of $3/4$-inch. Foams vary in composition, however, so fol-

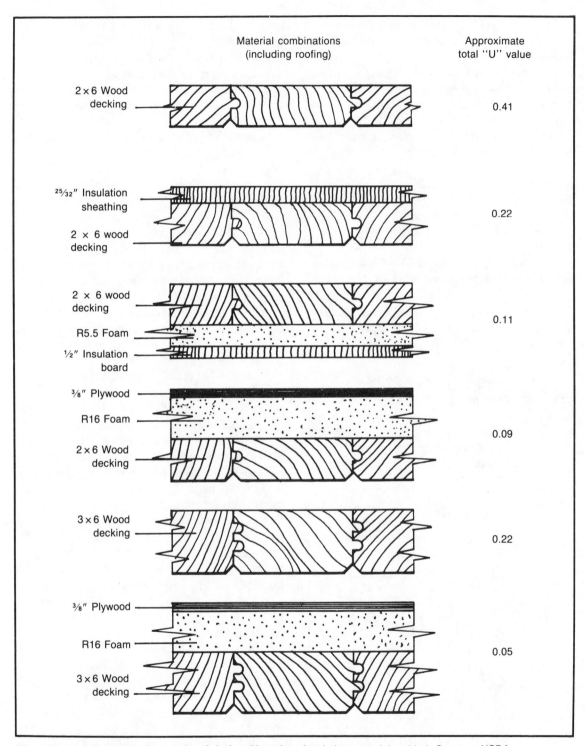

	Material combinations (including roofing)	Approximate total "U" value

2 × 6 Wood decking — 0.41

25/32" Insulation sheathing
2 × 6 wood decking — 0.22

2 × 6 wood decking
R5.5 Foam
½" Insulation board — 0.11

3/8" Plywood
R16 Foam
2 × 6 Wood decking — 0.09

3 × 6 Wood decking — 0.22

3/8" Plywood
R16 Foam
3 × 6 Wood decking — 0.05

Fig. 4-37. Insulating values for wood roof decks with various insulating materials added. Courtesy USDA

low manufacturer's instructions. Plywood at least ⅜-inch thick can be applied over the foam as a nailing base for shingles.

Trim for Low-Slope Roofs

Simple trim in the form of facia boards can be used at roof overhangs and at side and end walls. When 2-by-6-inch lookouts are used in the ceiling-beam roof, a 2-by-8-inch facia member is usually required to span the 48-inch spacing of the beams. In addition, a 1-by-2-inch facia molding may be added. Use two sixteenpenny galvanized nails in the ends of each lookout member. The facia molding may be nailed with six or sevenpenny galvanized nails on 16-inch centers.

Trim for roofs with nominal 2- or 3-inch-thick wood decking can consist of a 1-by-4- or 1-by-6-inch member with a 1-by-2-inch facia molding at the side and end-wall overhangs (Fig. 4-38). A 1-by-4-inch member can be used for 2-inch roof decking with or without the ²⁵⁄₃₂-inch insulating fiberboard. When 3-inch roof decking is used with the fiberboard, a

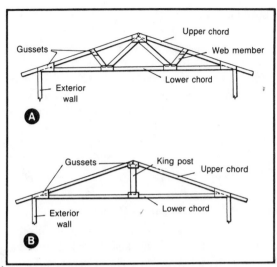

Fig. 4-39. Trussed rafters. A: W-type truss. B: King-post truss. Courtesy USDA

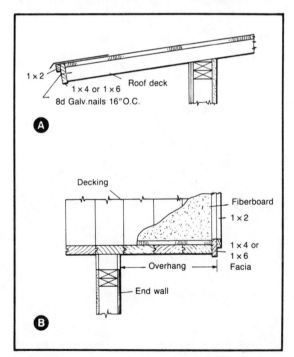

Fig. 4-38. Facia for wood-deck roof. A: Sidewall overhang. B: End-wall overhang. Courtesy USDA

1-by-6-inch piece is generally required. Nail the facia and molding to the decking with eightpenny galvanized nails spaced about 16-inches apart. The roof deck is now ready for the roofing material.

Pitched Roof

A pitched-roof house is commonly framed by one of two methods: with trussed rafters or with conventional rafter and ceiling joist members. These framing methods are used most often for roof slopes of 4-in-12 and greater. The common W-truss (Fig. 4-39A) for moderate spans requires less material than the joist and rafter system, because the members in the upper and lower chords are usually only 2 by 4 inches in size for spans of 24 to 32 feet. The king-post truss (Fig. 4-39B) for spans of 20 to 24 feet uses even less material than the W-truss, but is perhaps more suitable for light-to-moderate roof loads. Low-slope roof trusses usually require larger members. In addition to lowering material costs, the truss permits freedom in location of interior partitions, because only the sidewalls carry the ceiling and roof loads.

The roof sheathing, trim, roofing, interior ceiling finish, and type of ceiling insulation used do not vary a great deal between the truss and the conventional roof systems. For plywood or lumber sheathing, 24-inch spacing of trusses and rafters and joists is considered a normal maximum.

Greater spacing is acceptable, but usually requires a thicker roof sheathing, plus the application of wood stripping on the undersides of the ceiling joists and trusses to furnish a support for ceiling finish. Therefore, most W-trusses are designed for 24-inch spacing and joist-rafter construction for 24- or 16-inch spacing. Trusses generally require a higher grade dimension material than the joist and rafter roof.

Trussed Roof

The common truss or trussed rafter is most often fabricated in a shop. While some are constructed at the job site, an enclosed building provides better control for their assembly. These trusses are usually fastened together in one of three ways: with metal truss plates, with plywood gussets, or with ring connectors.

The metal truss plates, with or without prongs, are fastened in place on each side of member intersections. Some plates are nailed and others have supplemental nail fastening. Metal-plate trusses are usually purchased through a large lumber dealer or manufacturer and are not easily adapted to on-site fabrication.

The plywood-gusset truss may be nailed or nailed and glued. If on-site fabrication is necessary, the nailed gusset truss and the ring connector truss are probably the best choices. Many adhesives suitable for trusses generally require good temperature control and weather protection that are not usually available on site.

Completed trusses can be raised in place with a small mechanical lift on the top plates of exterior sidewalls. They can also be placed by hand over the exterior walls in an inverted position, and then rotated into an upright position. Mark the top plates of the two sidewalls for the location of each set of trusses. Trusses are fastened to the outside walls and to 1-by-4- or 1-by-6-inch temporary horizontal braces, which are used to space and align them until the roof sheathing has been applied. Locate these braces near the ridge line.

Trusses can be fastened to the top wall plates

by toenailing, but this is not always the most satisfactory method. The heel gusset in a plywood-gusset or metal-plate truss is located at the wall plate and makes toenailing difficult. However, two tenpenny nails at each side of the truss can be used in nailing the lower chord to the plate (Fig. 4-40A).

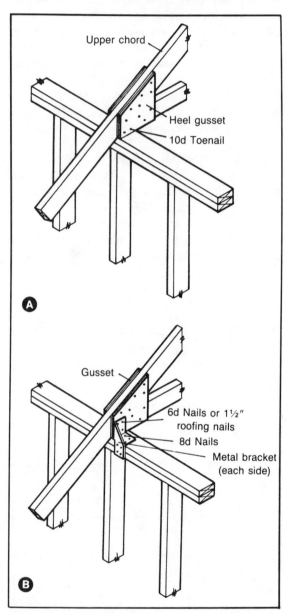

Fig. 4-40. Fastening trusses to wallplate. A: Toenailing. B: Metal connector. Courtesy USDA

Predrilling might be necessary to prevent splitting. A better system uses a simple metal connector or bracket obtained from local lumber dealers. Nail brackets to the wall plates at sides and top with eightpenny nails and to the lower chords of the truss with sixpenny or 1½-inch roofing nails (Fig. 4-40B), or as recommended by the manufacturer.

The gable-end walls for a pitched roof utilizing trusses are usually made the same way as those described in the previous section on "Wall Systems" and shown in Fig. 4-25.

Rafter and Ceiling Joist Roof

Conventional roof construction with ceiling joists and rafters (Fig. 4-41) can begin after all load-bearing and other partition walls are in place. The upper top plate of the exterior wall and the load-bearing interior wall serve as a fastening area for ceiling joists and rafters. Ceiling joists are installed along premarked exterior top wallplates and are toenailed to the plate with three eightpenny nails. The first joist is usually located next to the top plate of the end wall (Fig. 4-42). This provides edge-nailing for the ceiling finish. Ceiling joists crossing a center load-bearing wall are face-nailed to each other with three or four sixteenpenny nails. In addition, they are each toenailed to the plate with two eightpenny nails.

Angle cuts for the rafters at the ridge and at the exterior walls can be marked with a carpenter's square using a reference table showing the overall rafter lengths for various spans, roof slopes, and joist sizes. These tables can usually be obtained from your lumber dealer. If a rafter table is not available, lay out a baseline on the subfloor across the width of the house, marking an exact outline of the roof slope, ridge, board and exterior walls. Thus, a rafter pattern can be made, including cuts at the ridge, wall, and the overhang at the eaves.

Rafters are erected in pairs. First nail the ridge board to one rafter end with three tenpenny nails (Fig. 4-43). Then nail the opposing rafter to the first with a tenpenny nail at the top and two eightpenny nails toenailed at each side. The outside rafter is located flush with and a part of the gable-end walls.

While the ridge nailing is being done, toenail the rafters to the top plates of the exterior wall with two eightpenny nails. In addition, face-nail each rafter to the ceiling joist with three tenpenny nails. Install the remaining rafters the same way. When the ridge board must be spliced, it should be done at a rafter with nailing at each side.

If gable-end walls have not been erected with the end walls, the gable-end studs can now be cut and nailed in place. Toenail the studs to the plate

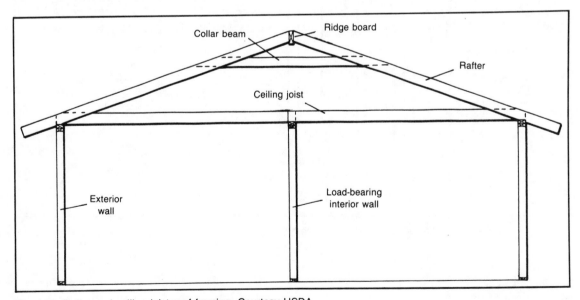

Fig. 4-41. Rafter and ceiling joist roof framing. Courtesy USDA

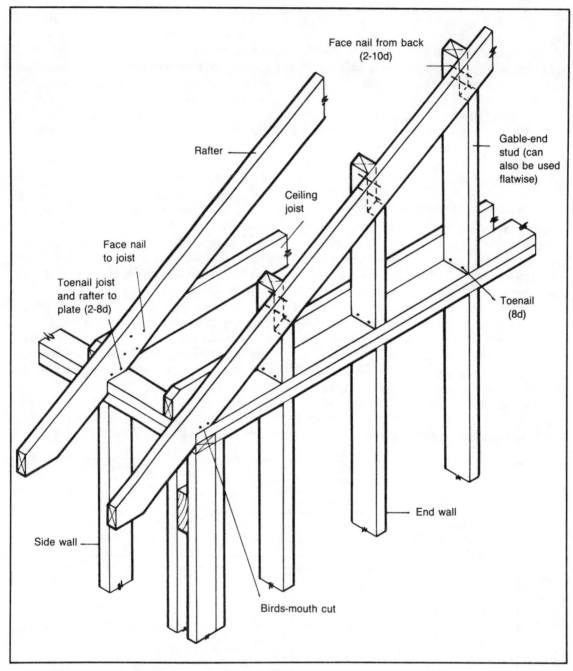

Face nail from back
(2-10d)

Rafter

Gable-end
stud (can
also be used
flatwise)

Ceiling
joist

Face nail
to joist

Toenail joist
and rafter to
plate (2-8d)

Toenail
(8d)

End wall

Side wall

Birds-mouth cut

Fig. 4-42. Fastening rafters and ceiling joists to plate and gable-end studs. Courtesy USDA

with eightpenny nails and face-nail them to the end rafter from the inside with two tenpenny nails. In addition, nail the first or edge ceiling joist to each gable-end stud with two tenpenny nails. Gable-end studs can also be used flatwise between the end rafter and top plate of the wall.

100

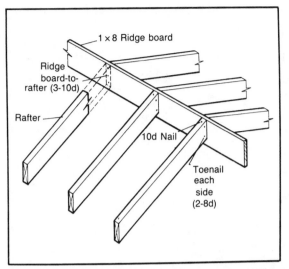

Fig. 4-43. Fastening rafters at the ridge. Courtesy USDA

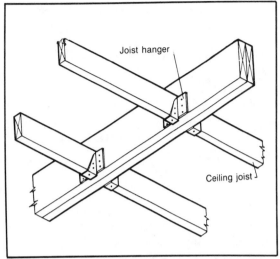

Fig. 4-44. Flush beam with joists hangers. Courtesy USDA

When the roof has a moderately low slope and the width of the house is 26 feet or greater, it is a good idea to nail a 1-by-6-inch collar beam to every second or third rafter (Fig. 4-41), using four eightpenny nails at each end.

Framing for Flush Living-Dining Area Ceiling

A living-dining-kitchen group is often designed as one open area with a flush ceiling throughout. This makes the rooms appear much larger than they actually are. When trusses are used, there is no problem, because they span from one exterior wall to the other. However, if ceiling joists and rafters are used, some type of beam is needed to support the interior ends of the ceiling joists. A flush beam, which spans from an interior cross wall to an exterior end wall, can be used for this. Fasten joists to the beams by means of joist hangers (Fig. 4-44). Nail these hangers to the beam with eightpenny nails and to the joist with sixpenny nails or 1½-inch roofing nails. Hangers are perhaps most easily fastened by nailing them to the end of the joist before the joist is raised in place.

An alternate method of framing is to tie a wood bracket at each pair of ceiling joists to a beam that spans the open living-dining area (Fig. 4-45). Block up and fasten this beam at each end at a height equal to the depth of the ceiling joists.

Roof Sheathing

Plywood or lumber roof sheathing is most commonly used for pitched roofs. Nominal 1-inch boards no wider than 8-inches can be used for trusses or rafters spaced not more than 24-inches on center. Sheathing (standard) and other grades of plywood are marked for the allowable spacing of the rafters and trusses for each species and thicknesses used. For example, a "24/0" mark indicates it is satisfactory as roof sheathing for 24-inch spacing of roof members, but not satisfactory for subfloor.

Lay up nominal 1-inch boards without spacing and nail them to each rafter with two eightpenny

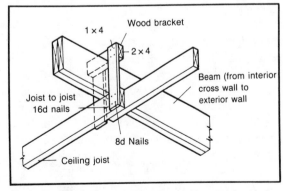

Fig. 4-45. Framing for flush ceiling with wood brackets. Courtesy USDA

nails. Lay plywood sheets across the roof members with staggered end joints. Use sixpenny nails for ⅜-inch and thinner plywood and eightpenny nails for ½-inch and thicker plywood. Space the nails 6 inches apart at the edges and 12 inches at intermediate fastening points.

When gable-end overhangs are used, extend the trim to the plywood or roofing boards when necessary before the 2-by-2 or 2-by-4-inch fly rafter (facia nailer) is nailed in place.

Roof Trim

Roof trim is installed before the roofing or shingles are applied. The cornice and gable (rake) trim for a pitched roof can be the same whether trusses or rafter-ceiling joist framing are used. In its simplest form, the trim consists of a facia board, sometimes with molding added. The facia is nailed to the ends of the rafter extensions or to the fly rafters at the gable overhang. With more complete trim, a soffit is usually included at the cornice and gables.

Cornice

The facia board at rafter ends or at the extension of the truss is often a 1-by-4- or 1-by-6-inch member (Fig. 4-46A). Nail the facia to the end of each rafter with two eightpenny galvanized nails. Trim rafter ends when necessary for a straight line. Nail 1-by-2-inch facia molding with one eightpenny galvanized nail at rafter locations. In an open cornice, a frieze board is often used between the rafters, serving to terminate siding or siding-sheathing combinations at the rafter line.

A simple closed cornice is shown in Fig. 4-46B. The soffit of plywood, hardboard, or other material is nailed directly to the underside of the rafter extensions. Blocking might be required between rafters at the wall line to serve as a nailing surface for the soffit. Use small galvanized nails in nailing the soffit to the rafters. When inlet attic ventilation is specified in the plans, it can be improved by a screen slot, or by small separate ventilators.

When a horizontal closed cornice is used, lookouts are fastened to the ends of the rafter and to the wall (Fig. 4-46C). They are face-nailed to the

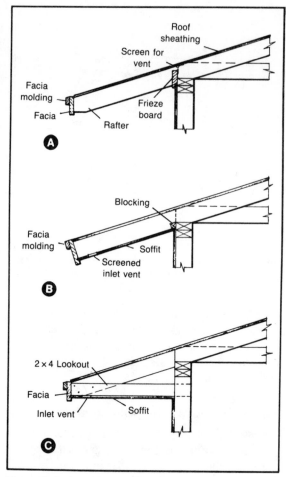

Fig. 4-46. Cornice trim. A: Open cornice. B: Sloped closed cornice. C: Horizontal closed cornice. Courtesy USDA

rafters and face- or toenailed to the studs at the wall. Use twelvepenny nails for the face-nailing and eightpenny nails for the toenailing.

Gable End

The gable-end trim may consist of a fly rafter, a facia board, and facia molding (Fig. 4-47A). The 2-by-2- or 2-by-4-inch fly rafter is fastened by nailing through the roof sheathing. Depending on the thickness of the sheathing, use sixpenny or eightpenny nails spaced 12 inches apart. In this type of gable end, the amount of extension should be governed by the thickness of the roof sheathing.

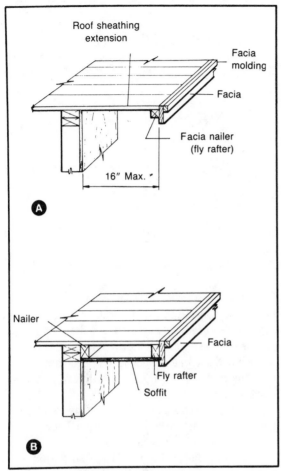

Fig. 4-47. Gable-end trim. A: Open gable overhang. B: Closed gable overhang. Courtesy USDA

When nominal 1-inch boards or plywood thicker than ½ inch are used, the extension should generally be no more than 16 inches. For thinner sheathings, limit the extensions to 12 inches.

A closed gable-end overhang requires nailing surfaces for the soffit. These are furnished by the fly rafter and a nailer or nailing blocks located against the end wall (Fig. 4-47B). An extension of 20 inches can be considered a limit for this type of overhang.

CHIMNEYS

Some type of chimney will be required for the heating unit, whether the home is heated by oil, gas, or solid fuel—and, of course, if you have any type of built-in or free-standing fireplace. The chimney is normally erected before the roofing is laid, but also can be installed afterwards. Chimneys should be structurally safe and provide sufficient draft for the heating unit and other utilities. Local building regulations often dictate the type to be used. A masonry chimney requires a stable foundation below the frost-line and construction with acceptable brick or other masonry units. Some type of flue lining is included, together with a cleanout door at the base.

A prefabricated chimney might cost less than the full masonry chimney, due to less cost in materials and labor. It might also provide a small saving in space. These chimneys are normally fastened to and supported by the ceiling joists and should come with a guarantee of safety. They are normally adapted to any type of fuel and come complete with roof flashing, cap assembly, mounting panel, piping, and chimney housing.

Framing for Chimneys

An inside chimney—whether it is of masonry or prefabricated—often requires that some type of framing be provided, if it extends through the roof. This might consist of simple headers between rafters below and above the chimney location, or require two additional rafter spaces (Fig. 4-48). The chimney should have a 2-inch clearance from the framing members and a 1-inch clearance from roof sheathing. When nominal 2- or 3-inch wood decking

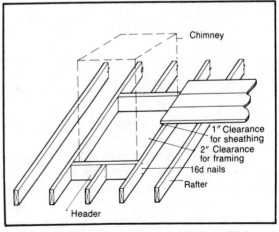

Fig. 4-48. Roof framing at chimney. Courtesy USDA

is used, a small header can be used at each end of the decking at the chimney location for support.

ROOF COVERINGS

Roof coverings should be installed soon after the cornice and rake trim are in place to provide protection for the remaining interior and exterior work. For the low-cost house, some of the most practical roof coverings are roll roofing or asphalt shingles, and fiberglass shingles. A good maintenance-free roof is important to protect the house and to prevent the additional cost involved in replacing a cheap roof after only a few years.

Asphalt Shingles

Under certain conditions, asphalt shingles may be used for roofs with slopes of 2-in-12 to 7-in-12 and steeper. The most common shingle is perhaps the 3-in-1, which is a 3-tab strip, 12 by 36 inches in size. The basic weight might vary somewhat, but the 235-pound (per square of 3 bundles) is now considered minimum. However, many roofs with 210-pound shingles are giving satisfactory service. A small gable-roof house uses about 10 squares of shingles, so using the better shingle would mean a minimal additional cost.

Installation is performed as follows: Use a single underlay of 15-pound saturated felt under the shingles for roof slopes of 4-in-12 to 7-in-12. A double underlay (double coverage) is required for slopes of 2-in-12 to 4-in-12. Roof slopes over 7-in-12 usually require no underlay.

For single underlay, start at the eave line with the 15-pound felt, roll across the roof, and nail or staple the felt in place as required. Allow a 2-inch head lap and install the second strip. This leaves a 34-inch exposure for the standard 36-inch-wide rolls. Continue in this manner.

A double underlay can be started with two layers at the eave line, flush with the facia board or molding. The second and remaining strips have 19-inch head laps with 17-inch exposure (Fig. 4-49). Cover the entire roof in this manner, making sure that all surfaces have double coverage. Use only enough staples or roofing nails to hold the underlay in place. Underlay is normally not required for wood shingles.

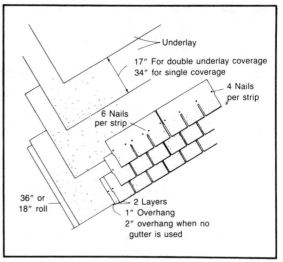

Fig. 4-49. Installing asphalt shingles. Courtesy USDA

Asphalt tab shingles are fastened in place with ¾- or ⅞-inch galvanized roofing nails or with staples, using at least four on each strip. Some roofers use six for each strip for greater wind resistance; one at each end and one at each side of each notch. Locate them above the notches so the next course covers them.

A starter strip and one or two layers of shingles are used at the eave line with a 1-inch overhang beyond the facia trim and ½- to ¾-inch extension at the gable end. When no gutters are used, the overhang should be about 2 inches. This will form a curve during warm weather for a natural drip. Metal edging or flashing is sometimes used at these areas. For slopes of 2-in-12 to 4-in-12, a 5-inch exposure can be used with the double underlay. For slopes of 4-in-12 and over, a 5-inch exposure may also be used with a single underlay.

Ridge

A Boston ridge is perhaps the most common method of treating the ridge portion of the roof. This consists of 12-by-12-inch sections cut from the 12-by-36 inch shingle strips. They are bent slightly and used in lap fashion over the ridge with a 5-inch exposure distance (Fig. 4-50). In cold weather, be careful that the sections do not crack in bending. The nails used at each side are covered by the lap of the next sec-

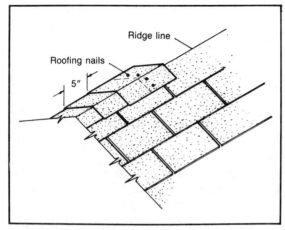

Fig. 4-50. Boston ridge. Courtesy USDA

tion. For a positive seal, use a small spot of asphalt cement under each exposed edge.

Roll Roofing

When cost is a factor in construction of a house, consider using mineral-surfaced roll roofing. While this type of roofing will not be as attractive as an asphalt shingle roof and perhaps not as durable, it can cost up to 15 percent less for a small house than standard asphalt shingles.

Roll roofing (65 pounds minimum weight in one-half lap rolls with a mineral surface) should be used over a double underlay coverage. Use a starter strip or a half-roll at the eave line with a 1-inch overhang and nail in place 3 to 4 inches above the edge of the facia (Fig. 4-51). When gutters are not included initially, use a 2-inch extension to form a drip edge. Space roofing nails about 6 inches apart. Surface nailing can be used when roof slopes are 5-in-12 and greater.

Place the second (full) roll along the eave line over a ribbon of asphalt roofing cement or lap-joint material. In low slopes, nail above the lap, apply cement, and position the next roll so that the nails are covered. Edge overhang should be ½ to ¾ inch at the gable ends. When vertical lap joints are required, nail the first edge, then use asphalt adhesive under a minimum 6-inch overlap. Use a sufficient amount of adhesive or lap-joint material to ensure a tight joint. On steep slopes, surface nailing along the vertical

edge is acceptable. The ridge can be finished with a Boston-type covering or by 12-inch-wide strips of the roll roofing, using at least 6 inches on each side.

Chimney Flashing

Flashing around the chimney at the junction with the roof is perhaps the most important flashed area in a simple gable roof. The Boston ridge over the shingles must be well installed to prevent wind-driven rain from entering, and the same is true for the flashing around the chimney. Prefabricated chimneys are supplied with built-in flashing that slides under the shingles above and over those below. A good caulking or asphalt sealing compound around the perimeter completes the installation.

A masonry chimney requires flashing around the perimeter, which is placed as shingle flashing under the shingles at sides and top, and extends at right angles up the sides (Fig. 4-52). In addition, counterflashing is used on the base of the chimney over the shingle flashing. This is turned in a masonry joint, wedged in place with lead plugs, and sealed with a caulking material. Galvanized sheet metal, aluminum, and terneplate (coated sheet iron or steel) are the most common types used for flashing around the chimney. If they are not rust-resistant, they should be given a coat or two of good metal paint.

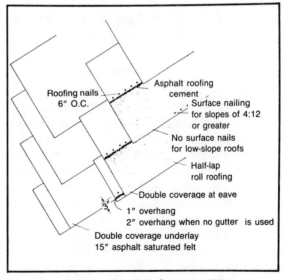

Fig. 4-51. Installing roll roofing. Courtesy USDA

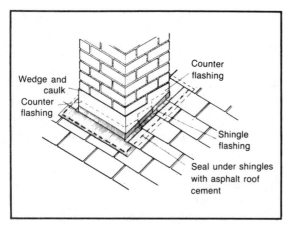

Fig. 4-52. Chimney flashing. Courtesy USDA

EXTERIOR WALL COVERINGS

Exterior coverings used over the wall framing commonly consist of a sheathing material followed by some type of finish siding. However, sheathing-siding materials (panel siding) serve as both sheathing and finish material. These materials are most often plywood or hardboards. While they are somewhat higher in price than sheathing alone, no other exterior covering material is needed.

Low-cost sheathing materials can be covered with various types of siding—from spaced vertical boards over plywood sheathing to horizontal bevel siding over fiberboard, plywood, or other types of sheathing. Study all combinations, taking into consideration cost, utility, and appearance.

Sheathing

In a low-cost house, it is advisable to use a sheathing or a panel-siding material that will provide resistance to racking and eliminate the need for diagonal corner bracing on the stud wall. Notching studs and installing bracing can add substantially to labor cost. When siding material does not provide this rigidity and strength, use some type of sheathing.

Materials that provide resistance to racking are: diagonal board sheathing, plywood, and structural insulation board (fiberboard) sheathing.

The fiberboard and plywood sheathing must be applied vertically in 4-by-8-foot or longer sheets. Edge and center nailing will provide the needed rack-ing resistance. Horizontal wood boards may also be used for sheathing but require some type of corner bracing.

Diagonal Boards

Diagonal wood sheathing should have a nominal thickness of ⅝ inch (resawn). Edges can be square, shiplapped, or tongue-and-groove. Widths up to 10 inches are satisfactory. Sheathing should be applied as near a 45-degree angle as possible (Fig. 4-53). Use three eightpenny nails for 6- and 8-inch-wide boards and four eightpenny nails for the 10-inch widths. Also provide nailing along the floor framing or beam faces. Make butt joints over a study unless the sheathing is matched, end and side. Depending on the type of siding used, sheathing should normally be carried down over the outside floor framing members. This provides an excellent tie between wall and floor framing.

Structural Insulating Board

Structural insulating board sheathing (fiberboard type) in 4-foot-wide sheets and in ²⁵⁄₃₂-inch regular-density, or ½-inch intermediate fiberboard grades provides the required rigidity without bracing. It must be applied vertically in 8-foot and longer sheets with edge and center nailing (Fig. 5-54). Space nails 3 inches apart along the edges and 6 inches apart at intermediate supports. Use 1¾-inch roofing nails for the ²⁵⁄₃₂-inch sheathing and 1½-inch nails for the ½-inch sheathing. Make vertical joints over studs. Siding is normally required over this type of sheathing.

Plywood

Plywood sheathing also requires vertical application of 4-foot-wide by 8-foot or longer sheets. Standard sheathing grade plywood is normally used for this purpose. Use ⁵⁄₁₆-inch minimum thickness for 16-inch stud spacing and ⅜-inch for 24-inch stud spacing. Space nails 6 inches apart at the edges and 12 inches apart at intermediate studs. Use sixpenny nails for ⁵⁄₁₆- or ⅜-inch. Because the plywood sheathing in ⁵⁄₁₆- or ⅜-inch sheets provides the necessary strength and rigidity, almost any type of siding can be applied over it.

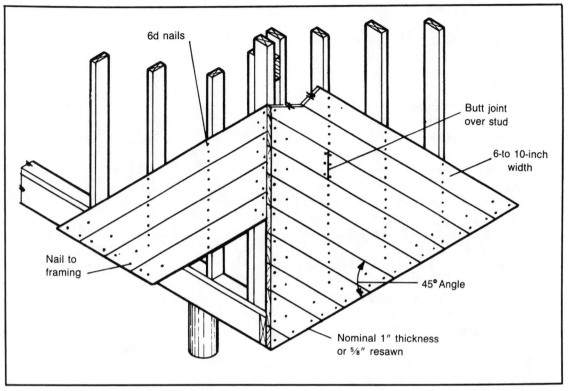

Fig. 4-53. Diagonal board sheathing. Courtesy USDA

Plywood and hardboard are also used as a single covering material without sheathing. Grades, Thickness, and types, however, vary from normal sheathing requirements.

Panel Siding

Large sheet materials for exterior coverage (panel siding) can be used alone and serve both as sheathing and siding. Plywood and hardboard are perhaps the most popular materials used for this purpose. The proper type and size of plywood and hardboard sheets with adequate nailing eliminate the need for bracing. Particleboard requires corner bracing.

These materials are quite reasonable in price. Also, plywood can be obtained grooved, rough-sawn, embossed, and with other surface variations, as well as in a paper-overlay form. Hardboard can also be obtained in a number of surface variations. The plywood surfaces are most suitable for pigmented stain finishes. The medium-density, paper-overlay ply-

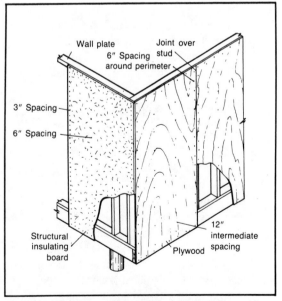

Fig. 4-54. Sheathing with insulating board or plywood (vertical application). Courtesy USDA

107

woods used for panel siding are normally exterior grades.

The thickness of plywood used for siding varies with the stud spacing. Grooved plywood, such as the "1-11" type, is normally ⅝-inch thick with ⅜-by-¼-inch deep grooves spaced 4 or 6 inches apart. This plywood is used when studs are spaced a maximum of 16 inches on center. Ungrooved plywoods should be at least ⅜ inch thick for 16-inch stud spacing and ½ inch thick for 24-inch stud spacing. Plywood panel siding should be nailed around the perimeter and at each intermediate stud. Use sixpenny galvanized siding or other rust-resistant nails for the ⅜-inch plywood and eightpenny for ½-inch and thicker plywood. Space 7 to 8 inches apart. Hardboard must be ¼ inch thick and used over 16-inch stud spacing. Exterior particleboard with corner bracing should be ⅝-inch thick for 16-inch stud spacing and ¾ inch thick for 24-inch stud spacing. Space nails 6 inches apart around the edges and 8 inches apart at intermediate studs.

The vertical joint treatment over the stud may consist of a shiplap joint as in the "1-11" paneling (Fig. 4-55A). This joint is nailed at each side after treating with a water-repellent preservative. When a square-edge butt joint is used, a sealant caulk should be used at the joint (Fig. 4-55B).

A square-edge butt joint may be covered with battens, which can also be placed over each stud as a decorative variation (Fig. 4-55C). Caulk joints and nail the batten over the joint with eightpenny galvanized nails spaced 12 inches apart. Nominal 1-by-2-inch battens are commonly used.

A good detail for this type of siding at gable ends consists of extending the bottom plate of the gable ⅝ inch to ¾ inch beyond the top of the wall below (Fig. 4-56). This allows a termination of the panel at the lower wall and a good drip section for the gable-end panel.

There are a number of sidings, mainly for horizontal application, which might be suitable for walls with or without sheathing. The types most suitable for use over sheathing are: the lower cost lap sidings of wood or hardboard, wood or other type shingles with single or double coursing, vertical boards, and several nonwood materials. Initial cost

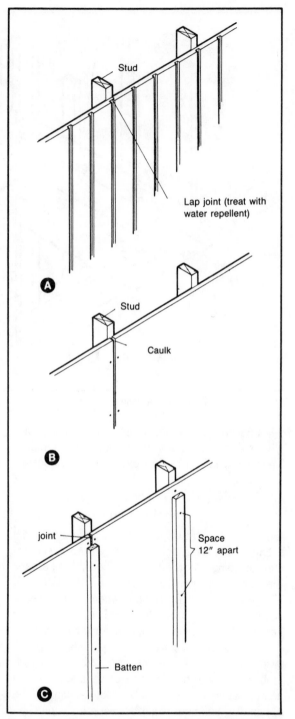

Fig. 4-55. Joint treatment for panel siding. A: Lap joint. B: Caulked butt joint. C: Butt joint with batten. Courtesy USDA

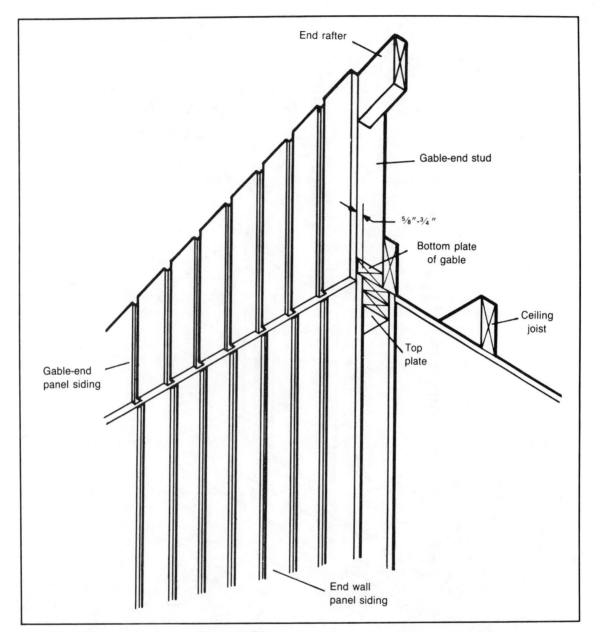

Fig. 4-56. Panel siding at gable end. Courtesy USDA

and maintenance should be the criteria in the selection. Drop siding and nominal 1-inch paneling materials can be used without sheathing under certain conditions. However, such sidings require a rigidly braced wall at each corner and a waterproof paper over the studs before application of the siding.

Application of Siding

Bevel Siding: When siding is used over sheathing, window and door frames are usually installed first. The exposed face of sidings such as bevel siding in ½-by-6-inch, ½-by-8-inch, or other sizes should be adjusted so that the butt edges coin-

109

cide with the bottom of the sill and the top of the drip cap of window frames (Fig. 4-57). Use a seven-penny galvanized siding nail or other corrosion-resistant nail at each stud crossing for the ½-inch-thick siding. Locate the nail so as to clear the top edge of the siding course below. Butt joints should be made over a stud. Install other horizontal sidings over sheathing in a similar manner. Nonwood sheathings require nailing into each stud.

Interior corners should butt against a 1½-by-1½-inch corner strip. This wood strip is nailed at interior corners before siding is installed. Exterior

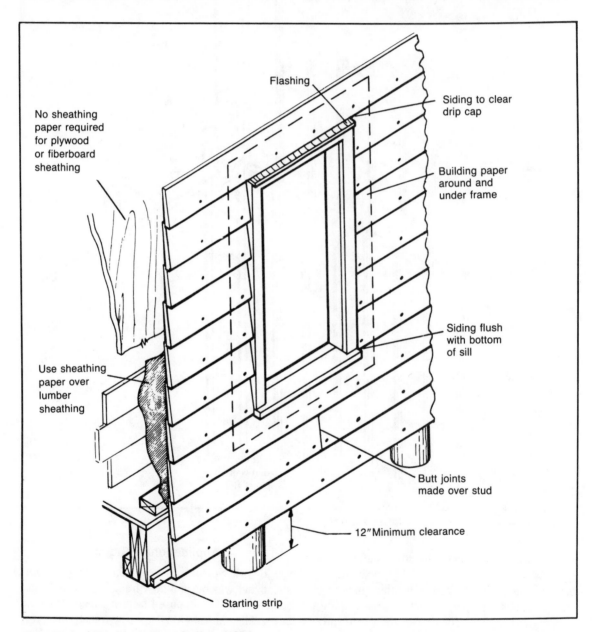

Fig. 4-57. Installing bevel siding. Courtesy USDA

110

corners can be mitered, butted against corner boards, or covered with metal corners. Perhaps the corner board and metal corner are the most satisfactory for bevel siding.

Vertical Siding: In low-cost house construction, some vertical sidings can be used over stud walls without sheathings; others require some type of sheathing as a backer or nailing base. Matched (tongue-and-groove) paneling boards can be used directly over the studs under certain conditions. First, some type of corner bracing is required for the stud wall. Second, nailers (blocking) between the studs are required for nailing points, and third, a waterproof paper should be placed over the studs (Fig. 4-58).

Blind-nail galvanized sevenpenny finish nails, which should be spaced no more than 24 inches apart vertically, through the tongue at each cross nailing block. When boards are nominal 6-inch and wider, face-nail an additional eightpenny galvanized nail. Boards should extend over and be nailed to the headers or stringers of the floor framing.

Rough-sawn vertical boards over a plywood backing provide a very good finish. The plywood should be an exterior grade or sheathing grade (standard) with exterior glue. It should be ½ or ⁵⁄₁₆ inch thick with nailing blocks between studs (Fig. 4-59). Rough-sawn boards 4 to 8 inches wide, sur-

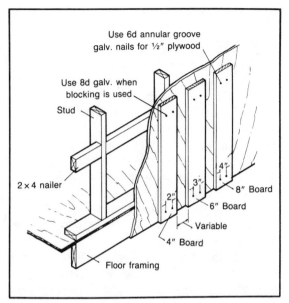

Fig. 4-59. Vertical boards over plywood. Courtesy USDA

faced on one side, can be spaced and nailed to the top and bottom wall plates, the floor framing members, and the nailing blocks. Use the surfaced side toward the plywood. A choice in the widths and the spacings of boards allows an interesting variation between houses.

EXTERIOR FRAMES

Windows and Doors

Windows and exterior door units, which include the frames as well as the sash or doors, are generally assembled in a manufacturing plant and arrive at the building site ready for installation. Doors might require fitting, however. Simple jambs and sill units for awning or hopper window sash can be made in a small shop with the proper equipment. The wood can then be treated with a water-repellent preservative, and the sash fitted and prehung. Only the sash would need to be purchased. However, this system of fabrication is practical only for the simplest units and only when a large number of the same type are required. If double-hung and sill units for awning or hopper window sashes are needed, it is desirable to select the lower-cost, standard-size units and use fewer windows. Using one large

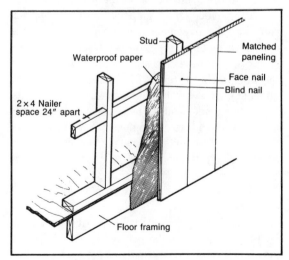

Fig. 4-58. Vertical paneling boards over studs. Courtesy USDA

111

double-hung window rather than two smaller ones will save money.

Window frames are generally made with nominal 1-inch jambs and 2-inch sills. Sashes for the most part are 1⅜-inch thick. Exterior door frames are made from 1½-to-1¾-inch stock. Exterior doors are 1¾ inches thick and the most common are the flush and the panel types.

As a general rule, the amount of natural light provided by the glass area in all rooms (except the kitchen) should be about 10 percent of the floor area. The kitchen can have natural or artificial light, but when an operating window is not available, ventilation should be provided. The same requirements apply to the bathroom.

From the standpoint of safety, houses should have two exterior doors (Fig. 4-60). Local regulations often specify any variations of these requirements. The main exterior door should be 3 feet wide and at least 6 feet, 6 inches high; 6 feet, 8 inches is a normal standard height for exterior doors. The service or rear door should be at least 2 feet, 6 inches wide; 2 feet, 8 inches is the usual width.

Perhaps the most common type of window used in houses is the double-hung unit (Fig. 4-61). It can be obtained in a number of sizes, is easily weatherstripped, and can be supplied with storms or screens. Frames are usually supplied with a prefitted sash, and with the exterior casing and drip cap in place.

Another type of window—which is quite reasonable in cost and perhaps the one most adaptable to small shop fabrication in a simple form—is the awning or hopper type (Fig. 4-62).

Other windows such as the casement sash and sliding units are also available, but generally their cost is somewhat greater than the two types described. The fixed or stationary sash can consist of a simple frame with the sash permanently fastened in place. The frame for the awning window would be suitable for this type of sash.

Perhaps the most practical exterior door, considering cost and performance, is the panel type. A number of styles and patterns are available, most of them featuring some type of glazed opening (Fig. 4-63). The solid-core flush door—usually more costly

than the panel type—should be used for exteriors in most central and northern areas of the country, in preference to the hollow-core type. A hollow-core door is ordinarily for interior use, because it warps excessively due to heat when used on the outside. For this reason it would even be unsatisfactory for exterior use in the southern areas.

Installation

Preassembled window frames are easily installed. They are made to be placed in the rough wall openings from the outside and are fastened in place with nails. When panel siding is used in place of a sheathing and siding combination, the frames are usually installed after the siding is in place. When horizontal siding is used with sheathing, the frames are fastened over the sheathing and the siding applied later.

To ensure a water-and windproof installation for a panel-siding exterior, place a ribbon of caulking sealant (rubber or similar base) over the siding at the location of side and head casing (Fig. 4-64). When using a siding material over the sheathing, use strips of 15-pound asphalt felt around the opening.

Place the frame in the opening over the caulking sealant (preferably with the sash in place to keep it square), and level the sill with a carpenter's level. Shims can be used on the inside if necessary.

After leveling the sill, check the side casing and jamb with the level and square. Now nail the frame in place using tenpenny galvanized nails through the casing and into the side studs and the header over the window (Fig. 5-65). While nailing, slide the sash up and down to see that it works freely. The nails should be spaced about 12 inches apart and both side and head casing fastened in the same manner. Other types of window units are installed similarly. When a panel siding is used, place a ribbon of caulking sealer at the junction of the siding and the sill. Follow this by installing a small molding such as quarter-round.

Door frames are also fastened over panel siding by nailing through the side and head casing. First cut and trim the header and joists (Fig. 4-66). Use a ribbon of caulking sealer under the casing. The top

112

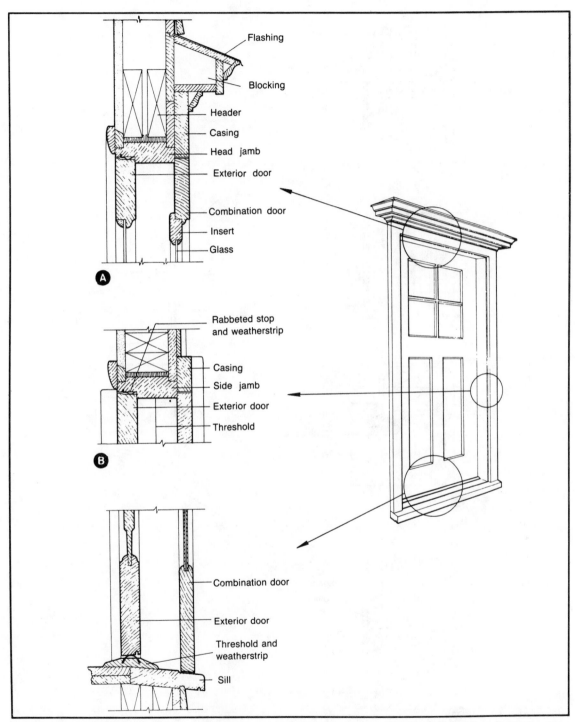

Fig. 4-60. Exterior door and combination door (screen and storm) cross sections. A: Head jamb. B: Side jam. C: Sill. Courtesy USDA

113

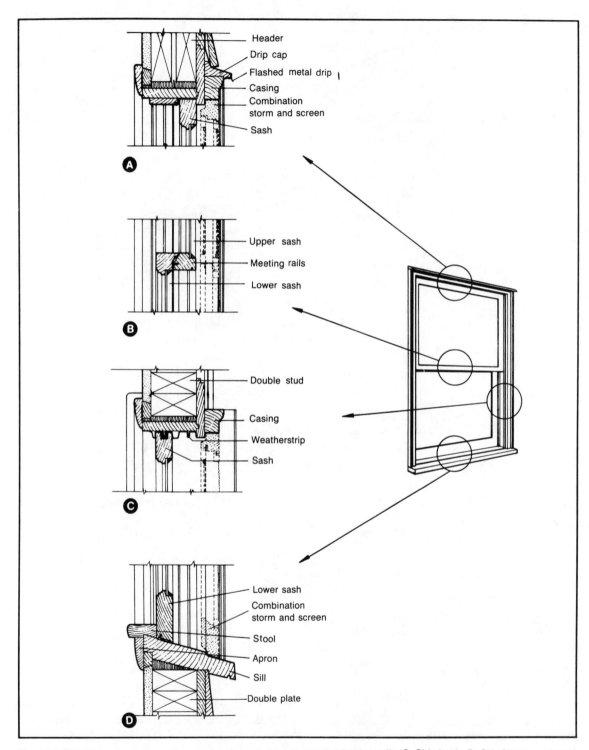

Fig. 4-61. Double-hung window unit, cross sections. A: Head jamb. B: Meeting rails. C: Side jamb. D: Sill. Courtesy USDA

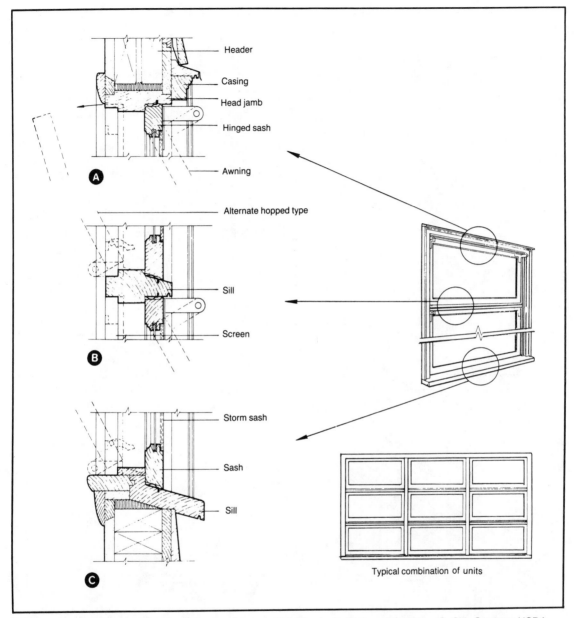

Fig. 4-62. Awning or hopper window, cross sections. A: Head jamb. B: Horizontal mullion. C: Sill. Courtesy USDA

of the sill should be the same height as the finish floor so that the threshold can be installed over the joint. The sill should be shimmed when necessary to have full bearing on the floor framing. Use quarter-round molding in combination with caulking when necessary for a tight, windproof joint. Use this under the door sill in conjunction with a panel siding or other single exterior covering. When joists are parallel to the plane of the door, headers and a short support member are necessary at the edge of the sill. The threshold is installed after the finish floor has been laid.

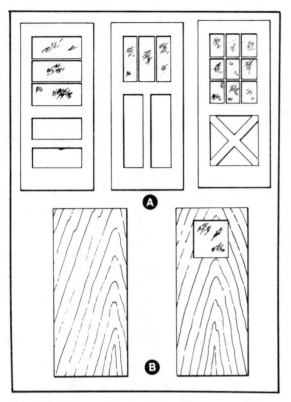

Fig. 4-63. Exterior doors. A: Panel type. B: Flush type. Courtesy USDA

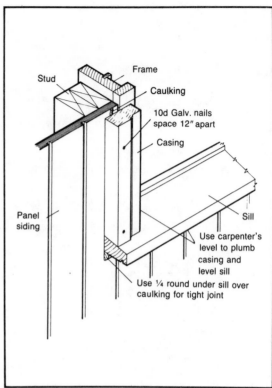

Fig. 4-65. Installation of double-hung window frame. Courtesy USDA

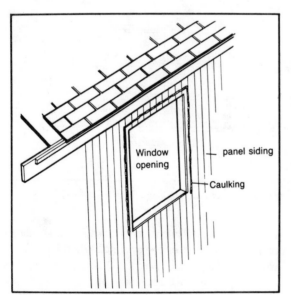

Fig. 4-64. Caulking around window opening before installing frame. Courtesy USDA

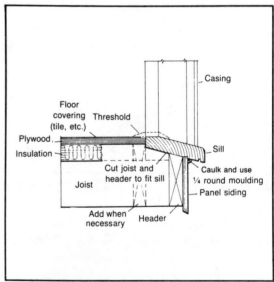

Fig. 4-66. Door installation at sill. Courtesy USDA

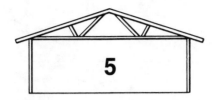

Insulation and Ventilation

Both insulation and ventilation are important factors in home construction. They make a home comfortable and save energy, resulting in lower utility bills. And insulating a new house is easily accomplished with just a few tools.

Thermal insulation is used in a house to minimize heat loss during the winter and to reduce the inflow of heat during the summer. Resistance to the passage of air is provided by materials used in wall, ceiling, and floor construction.

In constructing a crawlspace house, other factors besides insulation must also be considered. For example, protection from ground moisture by the use of a vapor barrier ground cover is necessary, especially if the crawlspace is enclosed with a full foundation or skirt boards. Attic and crawlspace ventilation might also be required.

The amount of insulation used in walls, floors, and ceilings usually depends on the geographic location of the house. In all parts of the country, insulation at least 3½ inches thick is essential in walls. Where winters are severe, the addition of foam sheathing or the alternative—a 2-by-6-inch stud wall with 5½-inch-thick insulation—is often necessary.

Most materials used in houses have some insulating value. Even air spaces between studs resist the passage of heat. However, when these stud spaces are filled or partially filled with a material highly resistant to heat transmission, namely insulation, the stud space has many times the insulating value of the air alone.

INSULATING MATERIALS

Commercial insulation is manufactured in a variety of forms and types, each with advantages depending on use (Fig. 5-1). Materials commonly used for insulation may be categorized as follows:

- Flexible, also known as blanket and batt
- Loose-fill
- Reflective
- Rigid
- Miscellaneous types.

Flexible insulation is manufactured in two types:

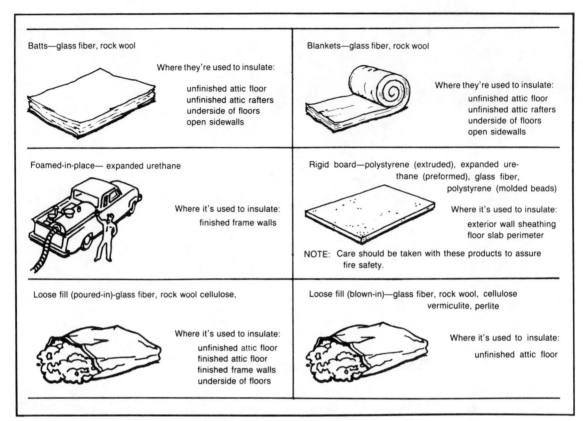

Batts—glass fiber, rock wool

Where they're used to insulate:

unfinished attic floor
unfinished attic rafters
underside of floors
open sidewalls

Blankets—glass fiber, rock wool

Where they're used to insulate:

unfinished attic floor
unfinished attic rafters
underside of floors
open sidewalls

Foamed-in-place— expanded urethane

Where it's used to insulate:

finished frame walls

Rigid board—polystyrene (extruded), expanded urethane (preformed), glass fiber, polystyrene (molded beads)

Where it's used to insulate:

exterior wall sheathing
floor slab perimeter

NOTE: Care should be taken with these products to assure fire safety.

Loose fill (poured-in)-glass fiber, rock wool cellulose,

Where it's used to insulate:

unfinished attic floor
finished attic floor
finished frame walls
underside of floors

Loose fill (blown-in)—glass fiber, rock wool, cellulose vermiculite, perlite

Where it's used to insulate:

unfinished attic floor

Fig. 5-1. Types of insulation. Courtesy USDA

blanket and batt. Blanket insulation is furnished in rolls or packages in widths suited to 16- and 24-inch studs and joint spacing. Usual thicknesses are 1½ to 2, and 3 inches. The body of the blanket is made of felted mats of mineral or vegetable fibers, such as rock or glass wool, wood fiber, and cotton. Organic insulation is treated to make it resistant to fire, decay, insects, and vermin. Most blanket insulation is covered with paper or other sheet material with tabs on the sides for fastening to studs or joists. One covering sheet serves as a vapor barrier to resist movement of water vapor and should always face the warm side of the wall. Aluminum foil, asphalt, or plastic laminated paper are commonly-used barrier materials (Fig. 5-2).

Batt insulation is also made of fibrous material, preformed to thicknesses of 4- and 6-inch for 16- and 24-inch joist spacing. It is supplied with or without a vapor barrier. One friction type of fibrous glass

batt is supplied without a covering and is designed to remain in place without the normal fastening methods (Fig. 5-3).

Loose fill is usually composed of materials used in bulk form, supplied in bags or bales, and placed by pouring, blowing, or packing by hand. The materials include rock or glass wool, wood fibers, shredded redwood bark, cork, wood pulp products, vermiculite, sawdust, and shavings.

Fill insulation is suited for use between first-floor ceiling joists in unheated attics. It is also used in sidewalls of existing houses that were not insulated during construction. Where no vapor barrier was installed during construction, suitable paint coatings, as described in this chapter, should be used for vapor barriers when blown insulation is added to an existing home.

Most materials reflect some radiant heat, and some materials have this property to a very high degree. Materials high in reflective properties include aluminum foil, sheet metal with tin coating, and paper products coated with a reflective oxide composition. Such materials can be used in enclosed stud spaces, in attics, and in similar locations to retard heat transfer by radiation. These reflective insulations are effective only when used where the reflective surface faces an air space ¾ inches or deeper. Where a reflective surface contacts another material, the reflective properties are lost and the material has little or no insulating value.

Reflective insulations are equally effective regardless of whether the reflective surface faces the warm or cold side. However, there is a difference in conductive ability and the resistance to heat flow, depending on the direction of heat flow (horizontal, up, or down) and the mean summer or winter temperature. It is more effective in the southern portion of the United States than in the northern portion.

Reflective insulation of the foil type is sometimes applied to blankets and to the stud-surface side of gypsum lath. Metal foil suitably mounted on some supporting base makes an excellent vapor barrier.

Rigid insulation is usually a fiberboard material manufactured in sheet and other forms. The most common types are made from processed wood, sugarcane, or other vegetable products. Structural insulating boards, in densities ranging from 15 to 31 pounds per cubic foot, are fabricated in such forms

Fig. 5-2. Laying in blanket insulation. Courtesy Owens Corning

Fig. 5-3. Laying batt insulation. Courtesy Owens Corning

as building boards, roof decking, sheathing, and wallboard. Although their primary purpose is structural, they have moderately good insulating properties.

Roof insulation is nonstructural and serves mainly to provide thermal resistance to heat flow in roofs. It is called "slab" or "block" insulation and is manufactured in rigid units ½ to 3 inches thick and usually 2 by 4 feet in size.

In house construction, perhaps the most common forms of rigid insulation are sheathing and decorative coverings in sheets or in squares. Sheathing board is made in thicknesses of ½ and ²⁵⁄₃₂ inch. It is coated or impregnated with an asphalt compound to provide water resistance. Sheets are made in a 2-by-8-foot size for horizontal applications and 4 by 8 feet or longer for vertical applications.

Some insulations do not fit in the classifications previously described. An example is insulation blankets made up of multiple layers of corrugated paper. Other types, such as lightweight vermiculite and perlite aggregates, are sometimes used in plaster as a means of reducing heat transmission.

Other materials are foamed-in-place insulation, which include sprayed and plastic foam types. Sprayed insulation is usually inorganic fibrous material blown against a clean surface that has been primed with an adhesive coating. It is often left exposed for acoustical as well as insulating properties.

Expanded polystyrene and urethane plastic foams may be molded or foamed in place. Urethane insulation may also be applied by spraying. Polystyrene and urethane in board form can be obtained in thicknesses from ½ to 2 inches.

Table 5-1 will provide insulating values of the various materials. These are expressed as "k" values or heat conductivity and are defined as the amount of heat, in British thermal units (BTUs), that will pass in 1 hour through 1 square foot of material

Table 5-1. Thermal Conductivity Values of Some Insulating Materials.

| Insulation group | | "k" range (conductivity) |
General	Specific type	
Flexible		0.25 — 0.27
Fill	Standard materials	.28 — .30
	Vermiculite	.45 — .48
Reflective (2 sides)		(1)
Rigid	Insulating fiberboard	.35 — .36
	Sheathing fiberboard	.42 — .55
Foam	Polystyrene	.19 — .29
	Urethane	.15 — .17
Wood	Low density	.60 — .65

[1] Insulating value is equal to slightly more than 1 inch of flexible insulation. (Resistance, "R" = 4.3)

Courtesy USDA

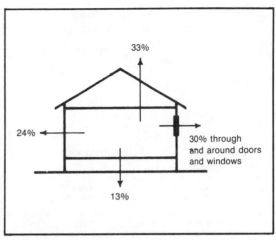

Fig. 5-4. Percentage and points of winter heat loss. Courtesy USDA

1 inch thick per 1 degree F temperature difference between faces of the material. Simply expressed, k-value represents heat loss, so the lower this numerical value, the better the insulating qualities.

Insulation is also rated on its resistance or "R" value, expressed as the total resistance of the wall or of a thick insulating blanket or batt. The "U" value is the overall heat-loss value of all materials in the wall. The lower this value, the better the insulating value.

FIX LOSS LEADERS FIRST

In the winter, about 30 percent of the heat loss in a home occurs through or around the doors and windows, 33 percent through the ceiling, 13 percent through the floor if there is a vented crawlspace or no basement, 24 percent through walls (Fig. 5-4). In the summer, heat enters through the ceiling, the side walls, and windows (Fig. 5-5). Therefore, for maximum year-round comfort, the side walls, floors, and ceiling need to be insulated in the form of weather-stripping, double glazing, or storm windows and doors.

In existing homes, first caulk and weatherstrip around windows and doors to eliminate air leaks. Next, insulate in the attic. Then insulate doors and windows by adding storm sashes of glass or plastic. Finally, if your home has a crawlspace, add under-floor insulation or insulate the walls of the crawlspace or basement. Table 5-2 provides insulating values of several insulating and building materials.

WHERE TO INSULATE

Place insulation on all outside walls and in the ceiling (Fig. 5-6A). In houses involving unheated crawl spaces, place it between the floor joists or along the perimeter of the wall. If a flexible type of insulation (blanket or batt) is used, it should be well-supported between joists by slats and a galvanized wire mesh, or by a rigid board with the vapor barrier installed toward the subflooring.

Press-fit or friction insulations fit tightly between joists and require only a small amount of support to hold them in place. Reflective insulation is often used for crawl spaces, but only one dead-air space should

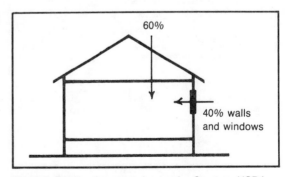

Fig. 5-5. Points of summer heat gain. Courtesy USDA

121

Table 5-2. Insulating Values of Insulating and Building Materials.

MATERIAL	INSULATION VALUE	
	"R" PER INCH THICKNESS	"R" FOR THICKNESS INDICATED
BATT or BLANKET INSULATION		
Wood or cellulose fiber with vapor barrier and paper facing	3.20-4.00	
Glass wool or mineral wool	3.00.3.80	
LOOSE FILL INSULATION		
Mineral wool (rock, glass, or slag)	2.80-3.70	
Vermiculite (expanded) Perlite (expanded)	2.13-2.70	
Cellulose	3.50-3.70	
RIGID INSULATION		
Polystyrene foam, extruded or expanded	4.00-5.40	
Polystyrene, molded beads	3.57	
Expanded urethane, sprayed or performed	5.80-8.00	
Polyurethane, expanded	6.25-8.00 +	
Glass fiber	4.00	
Insulating sheathing board (½" reg. density)	1.32	
(²⁵⁄₃₂" reg. density)	2.06	
CONSTRUCTION MATERIALS		
Concrete, sand, and stone aggregate	0.08	
Concrete block, three hole, 8"		0.95-1.11
Concrete block, lightweight aggregate, 8"		1.72-2.18
Concrete block, lightweight aggregate, 8" (Cores filled with vermiculite)		4.00-5.03
Face brick 4"		.44
Hardwoods, maple, oak, etc.	0.91	
Softwoods, fir, pine	1.25	
⅜" Plywood		0.47
½" Plywood		0.62
Hardboard, ¼" tempered		0.25
Wood siding, ½" thick clapboard		0.81
Asphalt shingles		.44
Aluminium or steel over flat sheathing		0.5-0.65
Gypsum or plaster board ⅜"		0.32
Gypsum or plaster board ½"		0.45
Plaster, brick or stucco	0.11-0.20	
Steel or aluminum	0.0007	
Glass	0.003	
DOORS		
Solid wood 1 inch	1.55	
Solid wood 2 inch		2.32
Solid wood 2 inch plus metal and glass storm door		3.45
WINDOWS (glass only)		
Single glazing		0.88
Double glazing (¼" to ½" air space)		1.60-1.75
Single glazing with storm windows		1.75-1.89
AIR SPACE		
Bounded by ordinary materials (vertical space)	¾" or more	.97
Horizontal-heat flow down	¾" or more	1.25
Horizontal-heat flow up	¾" or more	.85

Courtesy USDA

be assumed in calculating heat loss when the crawl space is ventilated. Place a ground cover of roll roofing or plastic film such as polyethylene on the soil of crawl spaces to decrease the moisture content of the space as well as of the wood members.

In 1½-story houses, insulation should be placed along all walls, floors, and ceilings that are adjacent to unheated areas (Fig. 5-6B). These include stairways, dwarf (knee) walls, and dormers. Provisions should be made for ventilation of the unheated areas.

Where attic space is unheated and a stairway is included, insulation should be used around the stairway as well as in the first-floor ceiling (Fig. 5-6C). It's a good idea to weatherstrip the door leading to the attic and to insulate walls adjoining an unheated garage or porch.

In houses with flat or low-pitched roofs, insulation should be used in the ceiling area with sufficient space allowed above for clear unobstructed ventilation between the joists. Insulation should be used along the perimeter of houses built on slabs, and a vapor barrier included under the slab.

In the summer, outside surfaces exposed to the direct rays of the sun may attain temperatures of 50 degrees F or more above shade temperatures and, of course, tend to transfer this heat toward the inside of the house. Insulation in the walls and attic areas retards the flow of heat so less heat is transferred through such areas.

Where air-conditioning systems are used, insulation should be placed in all exposed ceilings and walls in the same manner as insulating against cold-weather heat loss. Shading glass against direct rays of the sun and using insulated glass will aid in reducing the air-conditioning load.

Ventilation of attic and roof spaces is an important adjunct to insulation. Without ventilation, an attic space can become very hot and hold the heat for hours. Obviously more heat will be transmitted through the ceiling when the attic temperature is 150 degrees F than if it is 100 to 120 degrees F. Ventilation methods suggested for protection against cold-weather condensation also apply to protection against excessive hot-weather roof temperatures.

Storm windows or insulated glass will greatly reduce heat loss. Almost twice as much heat loss

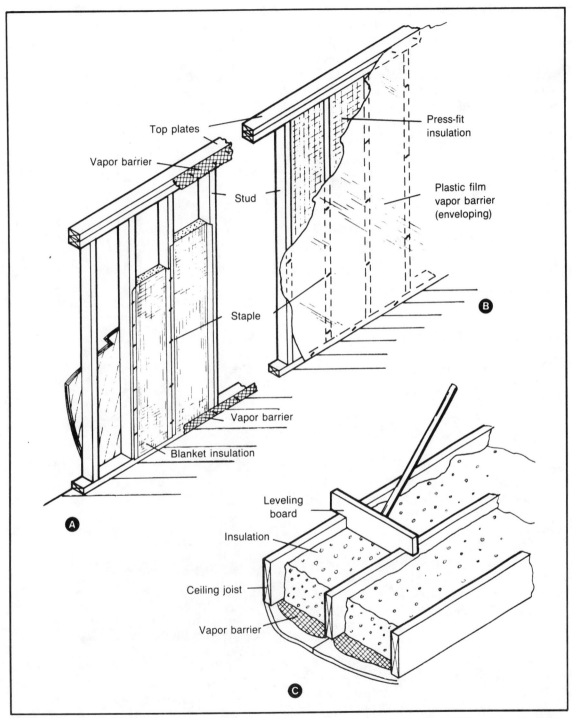

Fig. 5-6. Application of insulation. A: Wall section with blanket type. B: Wall section with "press-fit" insulation. C: Ceiling with full insulation. Courtesy USDA

occurs through a single glass as through window glazed with insulated glass or protected by a storm sash. Furthermore, double glass will normally prevent surface condensation and frost from forming on inner glass surfaces in winter. When excessive condensation persists, paint failures or even decay of the sash rail or other parts can occur.

INSTALLING INSULATION

Place blanket or batt insulation with a vapor barrier between framing members so that the tabs of the barrier lap the edge of the studs as well as the top and bottom plates. This method is often popular with contractors because it is less difficult than applying insulation to the drywall or rock lath (plaster base). To protect the head and soleplate as well as the headers over openings, it is good practice to use narrow strips of vapor barrier material along the top and bottom of the wall (Fig. 5-7A). Ordinarily, these areas are not covered well by the barrier on the blanket or batt. A hand stapler can be used to fasten the insulation and the barriers in place.

For insulation without a barrier (press-fit or friction-type), a plastic film vapor barrier such as 4-mil polyethylene is often used to envelop the entire exposed wall and ceiling (Fig. 5-7B). It covers the openings as well as window and door headers and edge studs. This system is one of the best for resisting vapor movement. Furthermore, it is not as inconvenient as stapling tabs of the insulation over the edges of the studs. After the drywall is installed or plastering is completed, trim the film around the window and door openings.

Place reflective insulation, in a single-sheet form with two reflective surfaces, in order to divide the space formed by the framing members into two approximately equal spaces. Some reflective insulations include air spaces and are furnished with nailing tabs. This type is fastened to the studs to provide at least a ¾-inch space on each side of the reflective surfaces.

Fill insulation is often used in ceiling areas and is poured or blown into place. Use a vapor barrier on the warm side (the bottom, in case of ceiling joists) before insulation is placed. A leveling board (shown in Fig. 5-6C) will allow a constant insulation

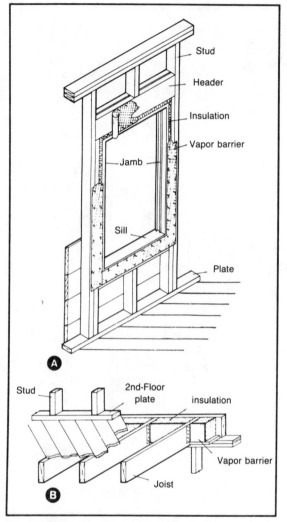

Fig. 5-7. Precautions in insulating. A: Around openings. B: Joist space in outside walls. Courtesy USDA

thickness. Thick batt insulation is also used in ceiling areas. Batt and fill insulation can also be combined to obtain the desired thickness. Place the vapor barrier against the back face of the ceiling finish. Ceiling insulation 6 or more inches thick greatly reduces heat loss in the winter and provides summertime protection.

PRECAUTIONS

Areas over door and window frames and along side and head jambs also require insulation. Because

124

these areas are filled with small sections of insulation, a vapor barrier must be used around the opening as well as over the header above the openings (Fig. 5-7A). Enveloping the entire wall eliminates the need for this type of vapor barrier insulation.

In 1½- and 2-story houses and in basements, the area at the joist header at outside walls should be insulated and protected with a vapor barrier (Fig. 5-7B).

Place insulation behind electrical outlet boxes and other utility connections in exposed walls to minimize condensation on cold surfaces.

VAPOR BARRIERS

Water vapor permeates most building materials. This presents problems because considerable water vapor is generated in a house from cooking, dishwashing, laundering, bathing, humidifiers, and other sources. During cold weather, this vapor can pass through wall and ceiling materials and condense in the wall or attic space. In severe cases, it could damage the exterior paint and interior finish, or even decay structural members. Vapor barrier, a material highly resistive to vapor transmission, should be used on the warm side of a wall or below the insulation in an attic space.

Among the effective vapor barrier materials are asphalt laminated papers, aluminum foil, and plastic films. Most types of blanket and batt insulation are provided with a vapor barrier on one side, some of them with paper-backed aluminum foil. Foil-backed gypsum lath or gypsum boards are also available and serve as excellent vapor barriers.

Some types of flexible blanket and batt insulations have a barrier material on one side. Attach such flexible insulations on the inside (narrow) edges of the studs. Cut the blanket long enough so that the cover sheet can lap over the face of the soleplate at the bottom and over the plate at the top of the stud space. (This, however, is not a common method of attachment.)

When a positive seal is desired, apply wall-height rolls of plastic film vapor barriers over studs, plates, and window and door headers. This system, called "enveloping," is used over insulation having no vapor barrier or to insure excellent protection when used over any type of insulation. The barrier should be fitted tightly around outlet boxes and sealed if necessary. A ribbon of sealing compound around an outlet or switch box will minimize vapor loss at this area. Cold-air returns in outside walls consisting of metal ducts prevent vapor loss and subsequent paint problems.

Paint coatings on plaster can be very effective as vapor barriers if materials are properly chosen and applied. They do not however, offer protection during the period of construction, and moisture might cause paint blisters on exterior paint before the interior paint can be applied. This is most likely to happen in buildings that are constructed when outdoor temperatures are 25 degrees F or more below inside temperatures. Paint coatings cannot be considered a substitute for the membrane types of vapor barriers, but they do provide some protection for houses where other types of vapor barriers were not installed during construction.

When painting, one coat of aluminum primer followed by two decorative coats of flat wall paint is effective. For rough plaster or for buildings in very cold climates, two coats of the aluminum primer might be necessary. A primer and sealer of the pigmented type, followed by decorative finish coats or two coats of rubber-base paint, are also effective in retarding vapor transmission.

Because no type of vapor barrier can be considered 100 percent resistive, some vapor leakage into the wall can be expected. Don't impede the flow of vapor to the outside by placing highly vapor-resistant materials on the cold side of the vapor barrier. For example, sheathing paper should be of a type that is waterproof but not highly vapor resistant. This also applies to "permanent" outer coverings or siding.

VENTILATION

Condensation of moisture vapor can occur in attic spaces and under flat roofs during cold weather. Even where vapor barriers are used, some vapor will probably work into these spaces around pipes and other inadequately protected areas, and some through the vapor barrier itself. Wood shingle and wood shake roofs do not resist vapor movement, while roofings as asphalt shingles and built-up roofs are highly resistant. The most practical method of removing and preventing the moisture is by adequately ventilating the roof spaces.

A warm attic that is inadequately ventilated and insulated can cause formation of ice dams at the cornice. After a heavy snowfall, heat causes the snow next to the roof to melt. Water running down the roof freezes on the colder surface of the cornice, forming an ice dam at the gutter. This could cause water to back up at the eaves and into the wall and ceiling. Similar dams often form in roof valleys. A well-insulated ceiling and adequate ventilation allows attic temperatures to be low and reduces melting of snow over the attic space.

In hot weather, ventilation of attic and roof spaces offers an effective means of removing hot air and thereby lowers the temperature in these spaces. Insulation between ceiling joists below the attic or roof space further retards heat flow into the rooms below. Louvered openings in the end walls of gable roofs are often installed for ventilation. Air movement through such openings depends primarily on wind direction and velocity. More positive air movement can be obtained by providing openings in the soffit areas of the roof overhang in addition to openings at the gable ends or ridge. Hip-roof houses are best ventilated by inlet ventilators in the soffit area and by outlet ventilators along the ridge. The differences in temperature between the attic and the outside will create an air movement when there is wind.

When there is a crawlspace under house or porch, ventilation is necessary to remove moisture vapor rising from the soil. Such vapor might otherwise condense on the wood below the floor and facilitate decay. A permanent vapor barrier on the soil of the crawlspace greatly reduces the amount of ventilating area required.

Tight construction (including storm window and storm doors) and the use of humidifiers create potential moisture problems. Avoid blocking ventilating areas.

AREA OF VENTILATORS

Types of ventilators and minimum recommended sizes have been generally established for various types of roofs. The minimum net area for attic or roof-space ventilators is based on the projected ceiling area of the rooms below (Fig. 5-8). The ratio of ventilator openings as shown are net areas, and the actual area must be increased to allow for any restrictions such as louvers and wire cloth or screen. The screen area should be double the specified net area shown in Figs. 5-8, 5-9, and 5-10.

To obtain extra area of screen without adding to the area of the vent, use a frame of required size to hold the screen away from the ventilator opening. Use as coarse a screen as conditions permit, not smaller than No. 16, because lint and dirt tend to clog fine-mesh screens. Install screens in such a way that paint brushes will not easily contact the screen and close the mesh.

Gable Roofs

Louvered openings are generally provided in the end walls of gable roofs and should be as close to the ridge as possible (Fig. 5-8A). The net area for the openings should be $\frac{1}{300}$ of the ceiling area (Fig. 5-9A). For example, where the ceiling area equals 1,200 square feet, the minimum total net area of the ventilators should be 4 square feet.

As previously explained, more positive air movement can be obtained if additional openings are provided in the soffit area. The minimum ventilation areas are shown in Fig. 5-9B.

Where there are rooms in the attic with sloping ceilings under the roof, the insulation should follow the roof slope and be placed so that there is a free opening of at least 1½ inch between the roof boards and insulation for air movement (Fig. 5-8C).

Hip Roofs

Hip roofs have air-inlet openings in the soffit area of the eaves and outlet openings at or near the peak. Minimum net areas of openings are shown in Fig. 5-9A. The most efficient type of inlet opening is the continuous slot, which should provide a free opening of not less than ¾ inch. The air-outlet opening near the peak can be a globe-type metal ventilator or several smaller roof ventilators located near the ridge. They can be located below the peak on the

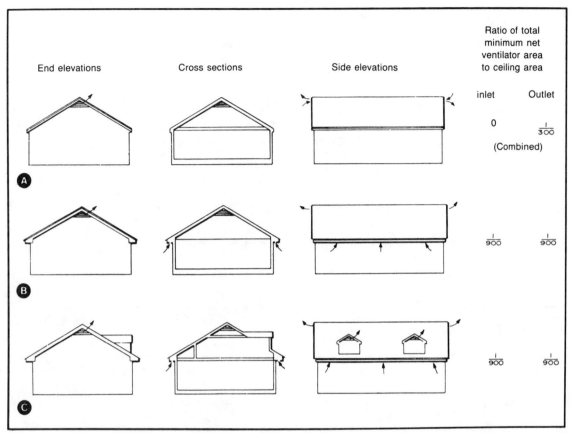

End elevations	Cross sections	Side elevations	Ratio of total minimum net ventilator area to ceiling area	
			inlet	Outlet
A			0	$\frac{1}{300}$
			(Combined)	
B			$\frac{1}{900}$	$\frac{1}{900}$
C			$\frac{1}{900}$	$\frac{1}{900}$

Fig. 5-8. Ventilating areas of gable roofs. A: Louvers in end walls. B: Louvers in end walls with additional openings in soffit area. C: Louvers at end walls with additional openings at eaves and dormers. Cross section of C shows free opening for air movement between roof boards and ceiling insulation of attic room. Courtesy USDA

rear slope of the roof so that they will not be visible from the front of the house. Gabled extensions of a hip-roof house are sometimes used to provide efficient outlet ventilators (Fig. 5-9B).

Flat Roofs

A greater ratio of ventilating area is required in some types of flat roofs than in pitched roofs because the air movement is less positive and more dependent upon wind. It is important that there be a clear open space above the ceiling insulation and below the roof sheathing for free air movement from inlet to outlet openings. Do not use solid blocking for bridging or for bracing over bearing partitions if its use prevents the air circulation.

Perhaps the most common type of flat or low-

pitched roof is one in which the rafters extend beyond the wall, forming an overhang (Fig. 5-10A). When soffits are used, this area can contain the combined inlet-outlet ventilators, preferably a continuous slot. When single ventilators are used, they should be distributed evenly along the overhang.

A parapet-type wall and flat roof combination may be constructed with the ceiling joists separate from the roof joists or combined. When members are separate, the space between can be used for an airway (Fig. 5-10B). Inlet and outlet vents are then located as shown, or a series of outlet stack vents can be used along the centerline of the roof in combination with the inlet vents. When ceiling joists and flat rafters are served by one member in parapet construction, vents may be located as shown in Fig.

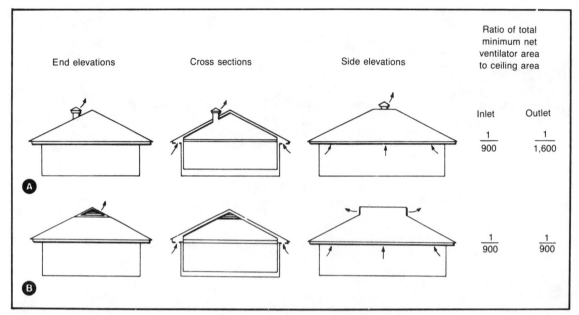

Fig. 5-9. Ventilating areas of hip roofs. A: Inlet openings beneath eaves and outlet vent near peak. B: Inlet openings beneath eaves and ridge outlets. Courtesy USDA

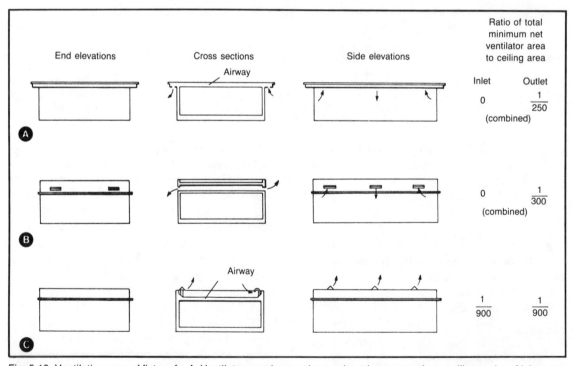

Fig. 5-10. Ventilating area of flat roofs. A: Ventilator openings under overhanging eaves where ceiling and roof joists are combined. B: For roof with a parapet where roof and ceiling joist are separate. C: For roof with a parapet where roof and ceiling joists are combined. Courtesy USDA

5-10C. Wall inlet ventilators combined with center steel outlet vents is another variable in this type of roof.

OUTLET VENTILATORS

Various styles of gable-end ventilators are available that are ready for installation. Many are made with metal louvers and frames, while others are made of wood to fit the house design more closely. However, the most important factors are to have sufficient net ventilating area and to locate ventilators as close to the ridge as possible without affecting house appearance.

One of the types commonly used fits the slope of the roof and is located near the ridge (Fig. 5-11A). It can be made of wood or metal; metal types are often adjustable to conform to the roof slope. A wood ventilator of this type is enclosed in a frame and placed in the rough opening much as a window frame (Fig. 5-11B). Other forms of gable-end ventilators which might be used are shown in Fig. 5-11C, D, and E.

A system of attic ventilation that can be used on houses with a wide roof overhang at the gable end consists of a series of small vents or a continuous slot located on the underside of the soffit areas (Fig. 5-11F). Several large openings located near the ridge might also be used. This system is especially desirable on low-pitched roofs where standard wall ventilators might not be suitable.

It is important that the roof framing at the wall line does not block off ventilation areas to the attic area. This might be accomplished by the use of a "ladder" frame extension. A flat nailing block used at the wall line will provide airways into the attic. This can also be adapted to narrower rake sections by providing ventilating areas to the attic (Fig. 5-12).

INLET VENTILATORS

Small, well-distributed ventilators or a continuous

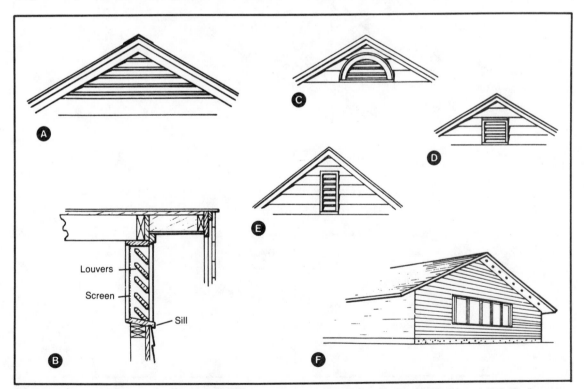

Fig. 5-11. Outlet ventilators. A: Triangular. B: Typical cross section. C: Half-circle. D: Square. E: Vertical. F: Soffit. Courtesy USDA

129

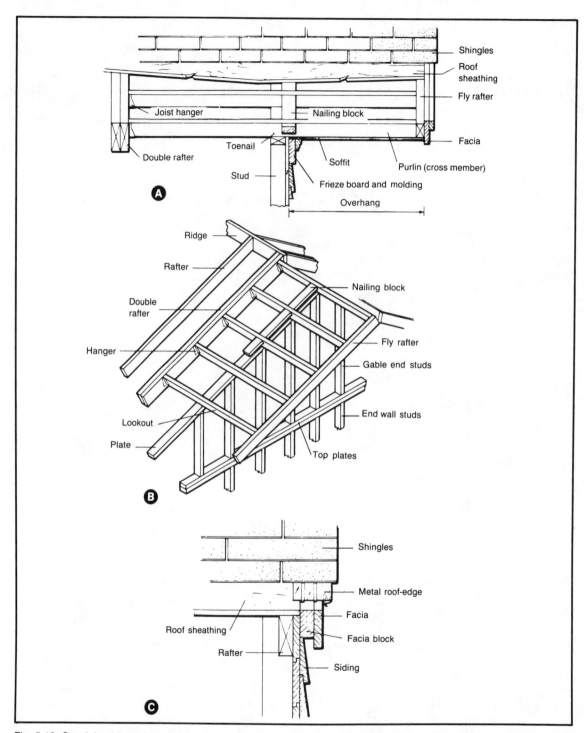

Fig. 5-12. Special gable-end extensions. A: Extra-wide overhang. B: Ladder framing for wide overhang. C: Close rake. Courtesy USDA

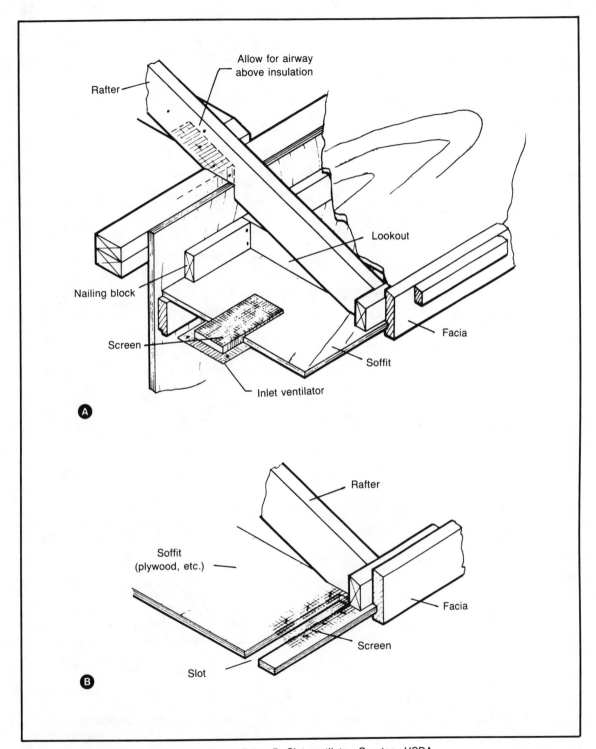

Fig. 5-13. Inlet ventilators. A: Small insert ventilator. B: Slot ventilator. Courtesy USDA

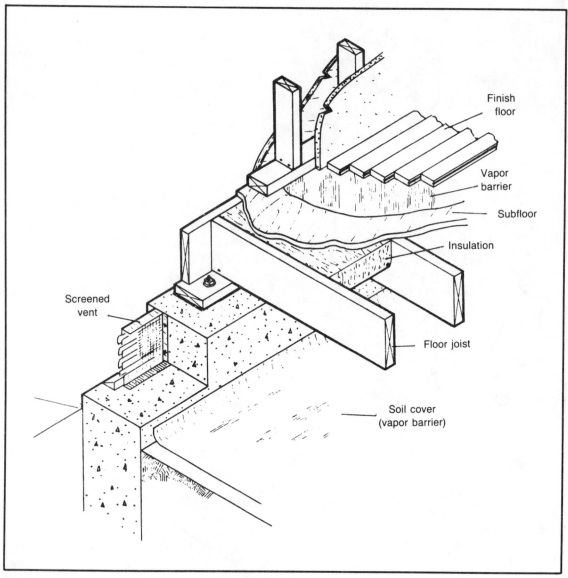

Fig. 5-14. Crawlspace ventilator and soil cover. Courtesy USDA

slot in the soffit provide inlet ventilation. These small louvered and screened vents can be obtained in most lumberyards or hardware stores and are simple to install.

Only small sections need to be cut out of the soffit; these can be sawed out before the soffit is installed. It is more desirable to use a number of smaller, well-distributed ventilators than several large ones (Fig. 5-13A). Any blocking that might be required between rafters at the wall line should be installed so as to provide an airway into the attic area.

A continuous screened slot, which is often desirable, should be located near the outer edge of the soffit near the facia (Fig. 5-13B). Locating the slot in this area will minimize the chance of snow entering. This type may also be used on the extension of flat roofs.

132

Crawlspace Ventilation and Soil Cover

The crawlspace below the floor of a basementless house and under porches should be ventilated and protected from ground moisture by the use of a soil cover (Fig. 5-14). The soil cover should be a vapor barrier with a perm value of less than 1.0. This includes such barrier materials as plastic films, roll roofing, and asphalt laminated paper. Such protection will minimize the effect of ground moisture on the wood framing members. High moisture content and humidity encourage staining and decay of untreated members.

Where there is a partial basement open to a crawlspace area, no wall vents are required if there is some type of operable window. The use of a soil cover in the crawlspace is still important, however. For crawlspaces with no basement area, provide at least four foundation wall vents near corners of the building. The total free (net) area of the ventilators should be equal to $1/160$ of the ground area when no soil cover is used. For a ground area of 1,200 square feet, a total net ventilating area of about 8 square feet is required, or 2 square feet for each of the four ventilators. More smaller ventilators having the same net ratio is satisfactory.

When using a vapor barrier ground cover, the required ventilating area is greatly reduced. The net ventilating area required with a ground cover is $1/1600$ of the ground area, or for the 1,200-square foot house, an area of 0.75 square feet. This should be divided between two small ventilators located on opposite sides of the crawlspace. Vents should be covered with a corrosion-resistant screen of No. 8 mesh.

The use of a ground cover is normally recommended under all conditions. It not only protects wood framing members from ground moisture but also allows the use of small, inconspicuous ventilators.

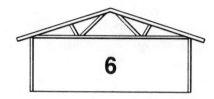

Interior Finishing

Interior finishing includes: interior walls and ceilings, windows and doors, floors and their various coverings. All of these should be considered carefully before you begin to build your house. Before shopping at your local lumber, hardware, carpeting, and tile stores to decide on the right finish for your individual needs, there are some basic facts that will help guide you.

WALLS AND CEILINGS

Some of the most practical low-cost materials for interior finish are the gypsum boards, the hardboards, the fiberboards, and plywood. These are usually prefinished and require only fastening to the studs and ceiling joists. Low-cost gypsum boards can also be obtained that have a paper facing, requiring painting or wallpapering.

A plaster finish is usually more costly than drywall. A plaster finish is also harder for the do-it-yourselfer to deal with; it takes more time and effort and many things can go wrong. It is recommended that you use drywall materials which are more easily installed. Then, except for paneling, there is only one more step: painting.

In addition to the four types of sheet materials mentioned, an insulating board or ceiling tile can be used for the ceiling. Prefinished tile is 12 by 12 inches. Installing larger sizes usually requires nailing strips fastened to the underside of the ceiling joists or truss members. Tile can also be applied to the underside of roof boards in a beam-type ceiling or on the inner face of the wood decking on a wood-deck roof.

Wood and fiberboard paneling in a tongue-and-groove, V-edge pattern of various widths can also be used as an interior finish—as an accent wall, for example. When applied vertically, nailers are used between or over the studs.

If your home is to be a low-cost first or second generation home, prices of materials are a big factor. Prefinished materials may cost a little more to purchase, but you will find, as a do-it-yourselfer, that they are easier to handle. You also end up with a more professional-looking job and will have less to do, so a little additional expense for prefinished interior finish materials is well worth the additional cost. Plus, the fact that you will be making a major saving on labor can allow you to spend a little more on finished materials of high quality. Before you begin to use any inside finishing materials, be sure that you have insulated well and that all of the wiring, heating ducts, and other utilities are roughed in.

The thickness of interior covering materials depends on the spacing of the studs or joists and the type of material. These requirements are usually a part of the working drawings or the specifications. However, for convenience, the recommended thickness for the various materials are listed in Table 6-1 based on their use and on the spacing of the fastening members.

For ceilings, when the long direction of the gypsum board sheet is at right angles to the ceiling joists, use ⅜-inch thickness for 16-inch spacing and ½-inch for 24-inch joist spacing. When sheets are parallel, spacing should not exceed 16 inches for ½-inch gypsum board. Fiberboard ceiling tile ½ inch thick requires 12-inch spacing of nailing strips.

Gypsum Board

Gypsum board sheets are normally 4 feet wide and 8 feet long, but can be obtained in lengths up to 16 feet. The edges along the length are usually tapered to allow for a filled and taped joint. This material can also be obtained with a foil back that serves as a vapor barrier on exterior walls. It is also available with vinyl or other prefinished surfaces. In new construction, ½-inch thickness is recommended for single-

Table 6-2. Gypsum Board Thickness (single layer).

Installed long direction of sheet	Minimum thickness	Maximum spacing of supports (on center)	
		Walls	Ceilings
	In.	In.	In.
Parallel to framing members	⅜	16	
	½	24	16
	⅝	24	16
Right angles to framing members	⅜	16	16
	½	24	24
	⅝	24	24

Courtesy USDA

layer application. In laminated two-ply application, two ⅜-inch sheets are used. The ⅜-inch thickness, while considered minimum for 16-inch stud spacing in single-layer applications, is normally specified for repair and remodeling work. Table 6-2 shows maximum member spacing for the various thicknesses of gypsum board.

To apply, start at one wall and proceed across the room. When using batt-type ceiling insulation, place it as each row of sheets is applied. Use fivepenny (1⅝-inch long) cooler-type nails for ½-inch gypsum, and fourpenny (1⅜-inch long) nails for ⅜-inch gypsum board. Ring-shank nails ⅛-inch shorter than these can also be used. Nailheads should be large enough so that they can't penetrate the surface.

Adjoining sheets should have only a light contact with each other. Stagger and Center end joints on a ceiling joist or bottom chord of a truss. One or two braces slightly longer than the height of the ceiling can be used to aid the installation of the gypsum sheets (Fig. 6-1). Nails are spaced 6 to 8 inches apart and should be very lightly dimpled with the hammerhead. Do not break the surface of the paper. Double-nail edge or end joints. Minimum edge nailing distance is ⅜ inch.

Wall-length sheets of gypsum board can eliminate vertical joints. The horizontal joint then requires only taping and treating. For normal application, horizontal joint reinforcing is not required. However, nailing blocks may be used between studs for a damage-resistant joint for the thinner gypsum sheets (Fig. 6-2A). Horizontal application is also suitable to

Table 6-1. Interior Material Finish Thickness.

Finish	Minimum material thickness (inches) when framing is spaced	
	16 inches	2.4 inches
Gypsum board	⅜	½
Plywood	¼	⅜
Hardboard	¼	
Fiberboard	½	¾
Wood paneling	⅜	½

Courtesy USDA

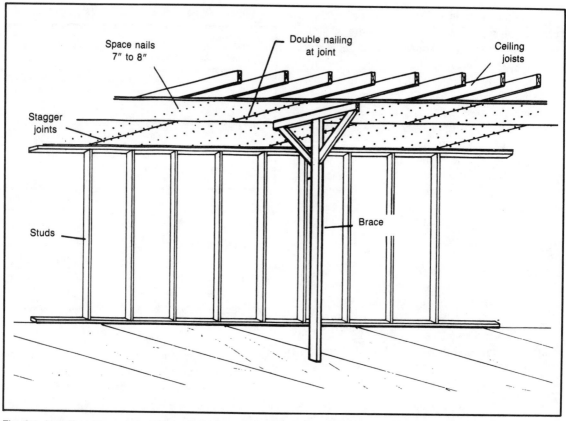

Fig. 6-1. Installing gypsum board on ceiling. Courtesy USDA

the laminated system in which ⅜-inch gypsum sheets are nailed vertically and room-length sheets are applied horizontally with a wallboard or contact adhesive. While this results in an excellent wall, it is much more costly than the single application.

Gypsum board applied vertically should be nailed around the perimeter and at intermediate studs with 1⅜- or 1⅝-inch nails, depending on the thickness. Space nails 6 to 8 inches apart. Joints should be made over the center of a stud with only light contact between adjoining sheets (Fig. 6-2B). Another method of fastening the sheets is called the "floating top." In this system, the top horizontal row of nails is eliminated and the top 6 or 8 inches of the sheet are free. This supposedly prevents fracture of the gypsum board when there is movement of the framing members.

The conventional method of preparing gypsum

drywall sheets for painting uses joint cement and perforated joint tape. Some gypsum board is supplied with a strip of joint paper along one edge, which is used in place of the tape. After the gypsum board has been installed and each nail driven in a "dimple" fashion (Fig. 6-3A), the walls and ceiling are ready for treatment. Joint cement ("spackle" compound), which comes in powder or ready-mixed form, should have a soft putty consistency so that it can be easily spread with a trowel or wide putty knife. Some manufacturers provide a joint cement and a finish joint compound that is more durable and less subject to shrinkage than standard fillers. The procedure for taping (Fig. 6-3B) is as follows:

1. Use a wide spackling knife (5 inches) to spread the cement over the tapered and other butt edges, starting at the top of the wall.

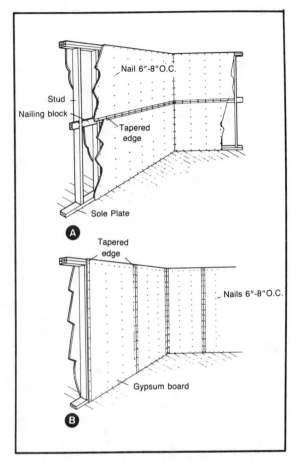

Fig. 6-2. Installing gypsum board on walls. A: Horizontal application. B: Vertical application. Courtesy USDA

2. Press the tape into the recess with the knife until the joint cement is forced through the small perforations.

3. Cover the tape with additional cement to a level surface, feathering the outer edges. When edges are not recessed, apply tape in the normal manner, but feather out the cement further so that the joint is level.

4. When dry, sand lightly and apply a thin second coat, feathering the edges again. A third coat might be required after the second has dried.

5. After cement is dry, sand smooth.

6. For hiding nail indentations at members between edges, fill with joint cement. A second coat is usually required.

Interior and exterior corners may be treated with perforated tape. Fold the tape down the center to a right angle (Fig. 6-3C). Apply cement on each side of the corner, then press tape in place with the spackle or putty knife. Finish with joint cement and sand when dry.

Wallboard corner beads of metal or plastic also can be used to serve as a corner guide and provide added strength. They are nailed to outside corners and treated with joint cement. The junction of the wall and ceiling can also be finished with a wood molding in any desired shape, which will eliminate the need for joint treatment (Fig. 6-3D). Use eight-penny finish nails spaced about 12 to 16 inches apart and nail them into the wallplate behind.

Treatment around window and door openings depends on the type of casing used. When a casing bead and trim are used instead of a wood casing, the jambs and the beads may be installed before or during application of the gypsum wall finish.

Plywood Hardboard

The application of prefinished 4-foot-wide hardboard and plywood sheets is relatively simple. They are normally used vertically and can be fastened with small finish nails (brads). Use nails 1½ inches long for ¼- or ⅜-inch thick materials, and space 8 to 10 inches apart at all edges and intermediate studs. Edge spacing should be about ⅜ inch. Set the nails slightly with a nail set. Many prefinished materials are furnished with small nails that require no setting because their heads match the color of the finish.

The use of panel and contact adhesives in applying prefinished sheet materials is becoming more popular and usually eliminates nails, except those used to align the sheets. Follow manufacturer's directions for this method of application.

In applying sheet materials such as hardboard or plywood paneling, it is good practice to insure dry, warm conditions before installing. Furthermore, place the sheets in an upright position against the wall, lined up approximately as they will be installed, and allow them to take on the condition of the room for at least 24 hours. This also applies for wood or fiberboard paneling.

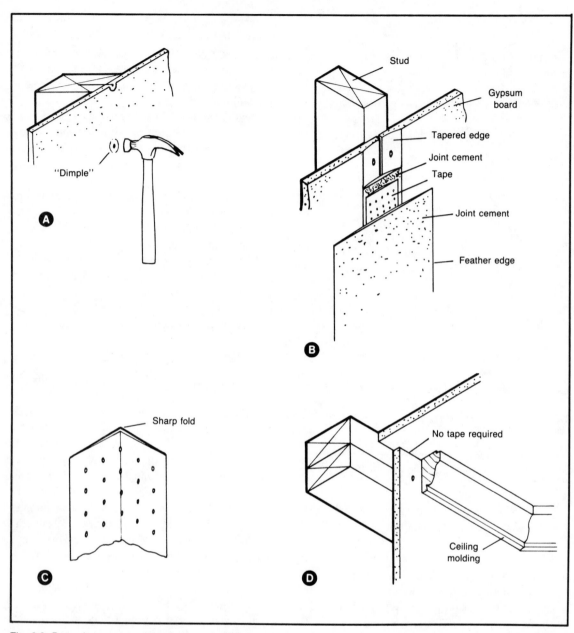

Fig. 6-3. Preparing gypsum drywall sheets for painting. A: Drive nails in "dimple" fashion. B: Detail of joint treatment. C: Corner tape. D: Ceiling molding. Courtesy USDA

Wood or Fiberboard Paneling

Tongue-and-groove wood or fiberboard (insulating board) paneling in various widths may be applied to walls in full lengths or above the wainscot. Wood paneling should not be too wide (nominal 8 inches) and should be installed at a moisture content of about 8 percent in most areas of the country. However, the moisture content should be about 6 percent in the dry Southwest and 11 percent in the southern and coastal areas of the country. In this type of application, use wood strips over the studs of nailing blocks placed between them (Fig. 6-4). Space the nailers not more than 24 inches apart.

For wood paneling, use a 1- to 2-inch finishing or casing nail and blind-nail through the tongue. For nominal 8-inch widths, a face nail may be used near the opposite edge. Fiberboard paneling (planking) is often supplied in 12- and 16-inch widths and is applied in the same manner as the wood paneling. In addition to the blind nail or staple at the tongue, two face nails might be required. A 2-inch finish nail is usually satisfactory, depending on the thickness. Panel and contact adhesives may also be used for this type of interior finish, eliminating the majority of the nails except those at the tongue. On outside walls, use a vapor barrier under the paneling.

Ceiling Tile

Ceiling tile can be installed in several ways, depending on the type of ceiling or roof construction.

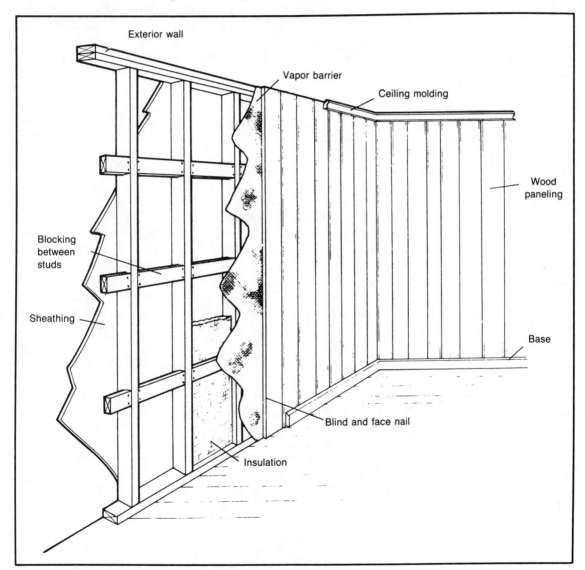

Fig. 6-4. Application of vertical paneling. Courtesy USDA

When a flat-surfaced backing is present, such as between beams of a beamed ceiling in a low-slope roof, the tiles are fastened with adhesive as recommended by the manufacturer. A small spot of mastic construction adhesive at each corner of a 12-by-12-inch tile is usually sufficient. When tile is edge-matched, stapling is also satisfactory.

A suspended ceiling with small metal or wood hangers that form supports for 24-by-48-inch or smaller drop-in panels is another system of applying this type of ceiling.

A third, and perhaps the most common, method of installing ceiling tile, is to use wood strips nailed across the ceiling joists or roof truss. There are spaced 12 inches on center. A nominal 1-by-3-inch or 1-by-4-inch wood member can be used for roof or ceiling members spaced not more than 24 inches on center (Fig. 6-5A). A nominal 2-by-2-inch or 2-by-3-inch member should be satisfactory for truss or ceiling joist spacing of up to 38 inches. Use two sevenpenny or eightpenny nails at each joist for the nominal 1-inch strip and two tenpenny nails for the nominal 2-inch strips. Use a low-density wood such as the softer pines, because most tile installation is done with staples.

When locating the strips, first measure the width of the room (the distance parallel to the direction of the ceiling joists). If, for example, this is 11 feet, 6 inches—use 10 full 12-inch-square tiles and a 9-inch-wide tile at each side edge. Thus, the second wood strips from each side are located so that they center the first row of tiles, which can now be ripped to a width of 9 inches. The last row will also be 9 inches, but do not rip these tiles until the last row is reached or they will not fit tightly. The tile can be fitted and arranged the same way for the ends of the room.

Ceiling tiles normally have a tongue on two adjacent sides and a groove on the opposite adjacent side. Start with the leading edge ahead and to the open side so that they can be stapled to the nailing strips. Use a small finish nail or adhesive at the edge of the tiles in the first row against the wall. Do the stapling at the leading edge and the side edge of each tile (Fig. 6-5B). Use one staple at each wood strip at the leading edge and two at the open side edge.

At the opposite wall, again use a small finish nail or adhesive to hold the tile in place.

Most ceiling tile of this type has a factory finish, and painting or refinishing is not required after it is in place. Because of this, take care not to soil the surface as it is installed. You can however, paint over the surface if you wish.

When installing ceiling tile, follow the manufacturer's instructions carefully. And before you do that, select your tiles carefully.

Most ceiling tiles are made of fiberboard; fine fibers are cut from wood or cane and then mixed with chemical binders and pressed into semi-hard, flat panels. Some tiles are made fire-resistant when special chemicals are added during this process. The most widely used tile size is 12 by 12 inches. They also come 12 by 24 inches, and many companies produce special feature sizes.

Tiles are made with tongue-and-groove edges for easy application. Plain, embossed, and patterned finishes are available, and some are molded with special texturing and square edges (instead of the common beveled edges) to make seams almost disappear when in place.

Acoustical tile is made from the same type of fiber, but undergoes an additional manufacturing process that helps it absorb much of the sound in a room. A well-designed acoustical tile can absorb up to 70 percent of the noise in a room.

To estimate the number of tiles you will need, do the following:

■ Measure ceiling length, then round it off to the next highest number.

■ Measure ceiling width, then round that off to the next highest number.

■ Add 1 to both measurements. (For example, if the ceiling measured 10 feet, 6 inches by 12 feet, 4 inches—round off both numbers and get 11 feet by 13 inches.

■ Add 1 to each number for a total of 12 feet by 14 inches.

■ Multiply those two numbers for the actual number of tiles needed (168 in this example).

■ Estimate the length of the wall molding by following the above instructions. Then measure

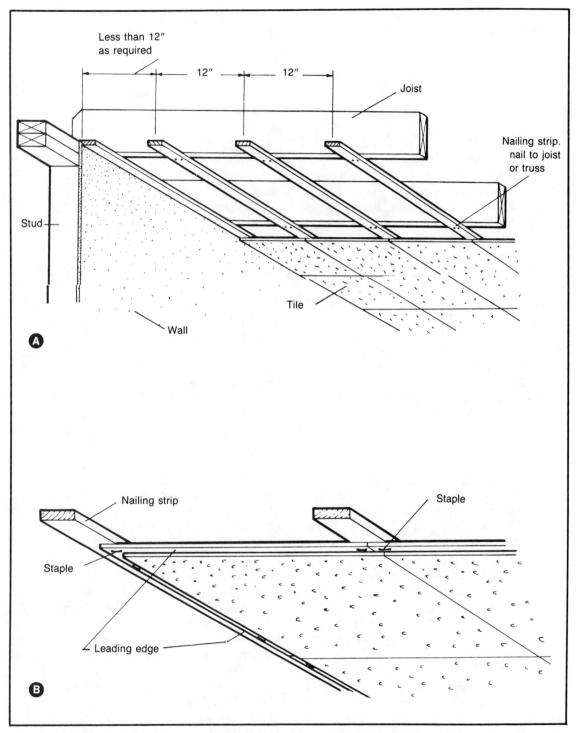

Less than 12″
as required

12″

12″

Joist

Nailing strip.
nail to joist
or truss

Stud

Tile

Wall

A

Nailing strip

Staple

Staple

Leading edge

B

Fig. 6-5. Ceiling tile installation. A: Nailing strip location. B: Stapling. Courtesy USDA

length and width of ceiling and round off to the next highest number.

■ Add the two and double.

■ If the ceiling measured 10 feet, 6 inches by 12 feet, 4 inches—round that off to 11 feet by 13 inches. Add the two numbers and double, for 48 feet of wall molding.

Tables 6-3 and 6-4 set out estimating various sizes of tile and furring strips.

If the ceiling is made of sound plaster, gypsum board, or other material that provides a sound, smooth, continuous backing, ceiling tiles can be applied with adhesive. If the ceiling has exposed joists, cracked plaster, or any other unsound surface, furring strips should be applied and the tiles nailed or stapled to the furring strips. (This would not be usual in new home construction.)

There are some basic rules for a neat job:

■ All cut tiles should be at the edges of the room, where the ceiling meets the walls.

■ Cut tiles at opposite ends of the room should be the same size.

■ Cut tiles should never be less than half a tile wide.

To determine width of border tiles, measure the total distance from wall to wall of the long side of the room. If the length comes out in exact feet, you do not need to cut any border tiles for that direction. If the distance does not come out in exact feet, add 12 to the number of inches remaining, and divide by two. The resulting number is the width of your border tile. Example: if the room is 10, feet 6 inches long—add 12 to the 6, divide 18 by 2 and the result is 9 inches. This is the proper width of your border. Adding 12 to the ''left over'' inches guarantees that your border tile will always be more than half the width of a full tile. Perform the same measurement for the short side of the room.

If you are not using furring strips, make sure that the tiles are applied with adhesive only to ceilings which are sound and even.

When using adhesives, surface preparation is important. Any painted surface should be checked carefully, because in some cases the paint might flake, peel, or become chalky, and tiles will not adhere. Test adhesive at different places around the room, and wait 48 hours to see how well the tiles adhere.

Using the technique outlined for determining the size of the border (above), make sure that border tiles will be the same on opposing sides of the room. Snap a chalk line along each side of the room the width of the border tile from the edge to align the first row of border tiles along both the short and long sides of the room.

Cut your first border tile to size. Because this tile fits into the corner, the dimensions of border tiles on both short and long sides of the room must be taken into consideration. For example, if your border tiles on the long sides of the room turn out to be 10 inches, and only the 9 inches on the short side of the room, the corner should be cut twice to make it 10 by 9 inches, allowing all other border tiles on both short and long sides of the room to line up properly with your full-size tiles.

Cut border tile on a flat surface, with the finished side up. Use a sharp knife or utility knife, and a clean straight-edge. Place the adhesive or cement in each corner of the tile (about an inch from the edge) and one daub of adhesive in the middle of the tile. Place the border tile into position in the corner.

It is important that the wide stapling edge lines up with the chalk marks on both sides for a solid fit, the flange must be exposed so the tongue of the next tile can slide into the tile already in place. If it is necessary to hold the tile while the adhesive dries, use a staple in each flange, but follow the adhesive manufacturer's instructions.

Place several border tiles in position along each edge, and then fill in between them with regular tiles. After completing the installation of the full-sized tile, measure and fit each border tile carefully on the opposite border. The installation of a border molding completes the job, giving you a neat, finished appearance.

If a ceiling has exposed joists (new construction), unsound plaster, or any uneven surface (existing construction), nail furring strips to the ceiling before tiles are applied.

Table 6-3. Estimating 12-by-12-Inch Tiles.

Room dimension perpendicular to joist direction

Room dimension parallel to joist direction	8'	9'	10'	11'	12'	13'	14'	15'	16'	17'	18'	19'	20'
8'	64	72	80	88	96	104	112	120	128	136	144	152	160
9'	72	81	90	99	108	117	126	135	144	153	162	171	180
10'	80	90	100	110	120	130	140	150	160	170	180	190	200
11'	88	99	110	121	132	143	154	165	176	187	198	209	220
12'	96	108	120	132	144	156	168	180	192	204	216	228	240
13'	104	117	130	143	156	169	182	195	208	221	234	247	260
14'	112	126	140	154	168	182	196	210	224	238	252	266	280
15'	120	135	150	165	180	195	210	225	240	255	270	285	300
16'	128	144	160	176	192	208	224	240	256	272	288	304	320
17'	136	153	170	187	204	221	238	255	272	289	306	323	340
18'	144	162	180	198	216	234	252	270	288	306	324	342	360
19'	152	171	190	209	228	247	266	285	304	323	342	361	380
20'	160	180	200	220	240	260	280	300	320	340	360	380	400

Courtesy Celotex Corp.

Table 6-4. Estimating Linear Footage of Furring Strips.

Room dimension perpendicular to joist direction

Room dimension parallel to joist direction	8'	9'	10'	11'	12'	13'	14'	15'	16'	17'	18'	19'	20'
8'	72	81	90	99	108	117	126	135	144	153	162	171	180
9'	80	90	100	110	120	130	140	150	160	170	180	190	200
10'	88	99	110	121	132	143	154	165	176	187	198	209	220
11'	96	108	120	132	144	156	168	180	192	204	216	228	240
12'	104	117	130	143	156	169	182	195	208	221	234	247	260
13'	112	126	140	154	168	182	196	210	224	238	252	266	280
14'	120	135	150	165	180	195	210	225	240	255	270	285	300
15'	128	144	160	176	192	208	224	240	256	272	288	304	320
16'	136	153	170	187	204	221	238	255	272	289	306	323	340
17'	144	162	180	198	216	234	252	270	288	306	324	342	360
18'	152	171	190	209	228	247	266	285	304	323	342	361	380
19'	160	180	200	220	240	260	280	300	320	340	360	380	400
20'	168	189	210	231	252	273	294	315	336	357	378	399	420

Courtesy Celotex Corp.

143

Seasoned 1-by-3-inch straight-grained soft woods such as pine, spruce, or fir make ideal furring strips. If the ceiling to which furring strips are to be applied has joists hidden by an existing ceiling, locate and mark these joists before the job is started. Joists can be located by driving a nail into the ceiling or by using a stud finder. Joists are usually 16 or 24 inches. After the first joist is found, measure across 16 inches and try again. After you have determined the spacing, locate and mark all joists with a chalk line so you can attach the furring strips without having to locate the joists again on each run.

Nail furring strips across the joists at right angles to the joists.

Attach the first furring strip on the ceiling immediately against the wall that runs at right angles to ceiling joists. Position the second furring strip so the distance between the center of the strip and the wall is the width of your border tile. The remaining furring strips must be exactly parallel to this strip, and the distance from center-to-center of each furring strip must be 12 inches. One of the easiest ways to position remaining furring strips is to cut a block of wood exactly 12 inches less the width of one furring strip and use this as a guide in positioning the remaining strips.

Use an eightpenny common nail for nailing the strips, with one nail at each joist.

All furring strips must be level. Use a long level to get a reading on all strips as they are added. Wood shims can be inserted between the joists and the furring strips for leveling if needed.

At the walls that run parallel to ceiling joists (at right angles to the furring strips), you may use scraps of furring to provide nailing or stapling position for border tile.

Now here are the instructions for stapling tiles to strips: First snap a chalk line along each side of the room to align the first row of border tiles along both the short and long sides of the room. This chalk line will run right down the center of the furring strip on one side, and across the furring strips on the other side.

Cut your first border tile to size. Again, because this tile will fit into a corner, dimensions of border tiles on both short and long sides of the room must be taken into consideration.

When cutting the border tiles, cut off the side without the wide stapling edge. The wide stapling flange must be exposed so the tongue of the next tile can fit into the groove of the tile already in place, guaranteeing a solid fit. Staple the tile in place, with three staples on the edge that is completely against a furring strip.

Place several border tiles in position along each edge, and then fill in between them with your full-size tiles. Measure and fit each tile carefully on the opposite border after you have worked your way across to the opposite wall.

Complete the job by using border molding, which also serves to hold the final border tiles in place where there is no flange left for stapling. At the border, where your access to the stapling area is limited, attach the tiles with small broad-headed nails, positioned as close to the wall as possible so that the finish molding will conceal them.

Ceiling tiles can be fitted around posts or pipes. Cut the tiles in half, each half following the contour of the pipe or post. Ceiling tiles can be fitted around ceiling fixture outlets or smaller pipes near the wall. When making a cut in a ceiling tile, always be sure to cut the tile face up, using a sharp utility knife.

FLOOR COVERINGS

Wood flooring is usually installed after the wall finish is applied. This is followed by installation of interior trim such as door jambs, casing, base, and other moldings. Wood floors can then be sanded and finished after the interior is completed. Some variation of this is necessary when casing bead and trim are used in drywall construction, in order to eliminate the need for standard wood casing around windows and doors. In this instance, the door jambs are installed before the wall finish is applied. Adjustment for the flooring thickness is then made by raising the bottom of the door jambs.

For resilient tile and prefinished wood block flooring, it is usually necessary to have most, if not all, the interior work completed before installation. This is especially true of resilient flooring. Manufacturers' recommendations usually state that all tradesmen, including painters, will have completed their work before the resilient tile is installed. In such cases, a resilient cove base might be a part of the finish, rather than the conventional wood base.

The term "finish flooring" usually applies to the material used as the final wearing surface. In its simplest form it may be merely a sealer or a paint finish applied to a tongue-and-groove plywood subfloor. Other floor coverings include linoleum, asphalt, rubber, vinyl, cork etc. These materials can vary a great deal.

The numerous flooring materials available may be used over a variety of floor systems. Choose the flooring for how well its properties fit its intended usage. Of the practical properties, perhaps durability and ease of maintenance are the most important. However, initial cost, comfort, and beauty or appearance must also be considered. Some areas call for special resistance to hard wear.

Hardwoods and softwoods are available as strip flooring in a variety of widths and thicknesses and as random-width planks and block flooring. Tile flooring is also available in a particleboard, which is manufactured with small wood particles combined with resin and fabricated under high pressure. Ceramic tile and carpeting are used in many areas not considered practical a few years ago. Plastic floor coverings used over concrete or stable wood subfloor are other variations.

Softwood finish flooring costs less than most hardwood species and is often used in bedroom and other areas where traffic is light. It is less dense than the hardwoods, less wear-resistant, and shows surface abrasions more readily. Softwoods most commonly used for flooring are southern pine, Douglas fir, redwood, and western hemlock.

Table 6-5 lists the grades and descriptions of softwood strip flooring. Softwood flooring has tongue-and-groove edges and may be hollow-backed or grooved. Some types are also end-matched. Vertical-grain flooring generally has better wearing qualities than flat-grain flooring under hard usage.

Hardwood strip flooring in the best grades is used in many high-cost houses (Fig. 6-6). Thinner tongue-and-groove strip flooring and thin square-edge flooring are lower in cost than $25/32$-inch strip flooring, and might be considered for use initially or at a later date. The use of low-grade softwood strip flooring with a natural or painted finish can also be considered. However, the installation costs of these materials are usually higher than those for resilient coverings. Many types require sanding and finishing after they are nailed in place. Another wood floor-

Table 6-5. Grade and Description of Strip Flooring.

Species	Grain orientation	Size		First grade	Second grade	Third grade
		Thickness	Width			
		In.	In.			
		SOFTWOODS				
Douglas-fir and hemlock	Edge grain	3/4	1⅛-5⅛	B and Better	C	D
	Flat grain	3/4	1⅛-5⅛			
Southern pine	Edge grain and Flat grain	5/16-1¼	1⅛-5⅛	B and Better	C,C and Better	D (and No.2)
		HARDWOODS				
Oak	Edge grain	3/4	1½-3¼	Clear	Select	-----------------
	Flat grain	11/32	1½, 2			
		15/32	1½, 2	Clear	Select	No. 1 Common
Beech, birch, maple, and pecan[1]		25/32	1½-3¼	First grade	Second grade	Third grade
		3/8	1½, 2¼			
		1/2	1½, 2¼			

[1]Special grades are available in which uniformity of color is a requirement.

Courtesy USDA

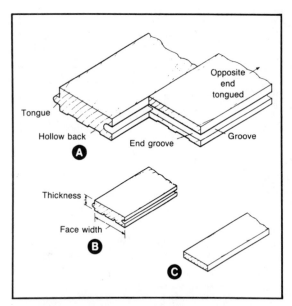

Fig. 6-6. Strip flooring. A: Side and end matched. B: Side matched. C: Square-edged. Courtesy USDA

½ to ⅝ inch away from the wall. Face-nail so that the base and shoe will cover the space and nails (Fig. 6-8B). Next, blind-nail the first strip of flooring and each subsequent piece with eightpenny flooring nails (for 2/32-inch thickness) at each joist crossing. Nail through the tongue and into the joist below with a nail angle of 45 degrees. Set each nail to the surface of the tongue. Each piece should be driven up lightly for full contact. Use a hammer and a short piece of scrap flooring to protect the edge. The last flooring strip must be face-nailed at the wall line, again allowing ½ to ⅝ inch space for expansion of the flooring.

Other thinner types of strip flooring are nailed in the same way, except that sixpenny flooring nails will probably be sufficient. Square-edge flooring must be face-nailed using two 1½-inch finish nails (brads) on about 12-inch centers.

Wood Block Flooring: Wood block flooring with matching tongue-and-groove edges is often fastened by nailing, but may be installed with an adhesive. Other block flooring made from wood-base materials are available. Manufacturers of these specialty floors can supply the correct adhesive as well as instructions for laying. Their recommended methods are based on years of experience, and when

ing, the parquet or wood block floor (Fig. 6-7), is usually prefinished. It is supplied in squares, and is installed by nailing or with an adhesive. These materials, while costing more than strip flooring, require no finishing. The choice is up to the individual budget and decorating requirements.

Installing Flooring

Strip Flooring: Before laying strip flooring, be sure that the subfloor is clean. When board subfloor is used, it should be covered with a building paper. This will aid in preventing air infiltration and help maintain a comfortable temperature at the floor level when a crawlspace is used. Chalkline the building paper at the joists as a guide in nailing the strip flooring. Before laying the flooring, open the bundles and spread out and expose the flooring to a warm, dry condition for at least 24 hours, preferably 48 hours. Moisture content of the flooring should be 6 percent for dry southwest areas, 10 percent for southern and coastal areas, and 7 percent for the remainder of the country. Your local lumber dealer can help you with this information if you are uncertain.

Lay strip flooring at right angles to the joists (Fig. 6-8A). Start at one wall, placing the first board

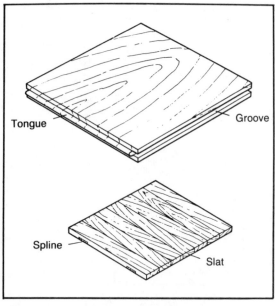

Fig. 6-7. Wood block flooring. Courtesy USDA

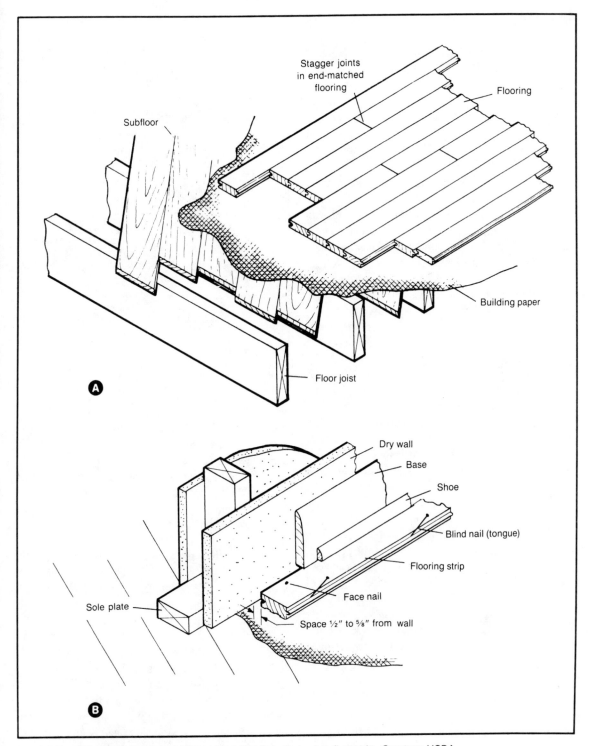

Fig. 6-8. Installing strip flooring. A: General application. B: Laying first strip. Courtesy USDA

followed should provide years of maintenance-free service.

Resilient Tile: Most resilient tiles are applied in the same way. The underlayment may be plywood, which serves both as a subfloor and an underlayment for the tile. A subfloor of wood boards requires an underlayment of plywood, hardboard, or particleboard. Drive nails flush with the surface, fill cracks and joints and sand smooth, then thoroughly clean the surface.

Mark a center baseline on the subfloor in each direction of the room. Centerlines should be exactly 90 degrees (a right angle) with each other. This can be assured by using a 3:4:5-foot measurement along two sides of and the diagonal. In a large room, a 6:8:10-foot measuring combination can be used (Fig. 6-9A).

Spread the adhesive with a serrated trowel (as recommended by the manufacturer) over one of the quarter-sections outlined by the centerlines. Waiting (drying) time should conform to the directions for the adhesives.

Starting at one inside corner, lay the first tile exactly in line with the marked centerlines. Lay the second tile adjacent to the first on one side (Fig. 6-9B). Lay the third tile adjacent to the first at the other centerline on the other side of the quarter-section. Thus, in checkerboard fashion, the entire section can be covered. Cover the remaining three sections in the same way. Some tile only needs to be pressed in place; others should be rolled after installing for better adherence.

The edge tiles around the perimeter of the room must be trimmed to fit to the edge of the wall. Allow a clearance of ⅛ to ¼ inch at all sides for expansion. This edge is covered with a cove base of the same resilient material or with a standard wood base. Wood base is usually lower in cost than the resilient cove base, but installation costs are somewhat greater (unless, of course, you are doing it yourself).

DOORS AND TRIM

The rough door openings provided during framing of the interior walls should accommodate the assembled door frames. The allowance was 2½ inches plus

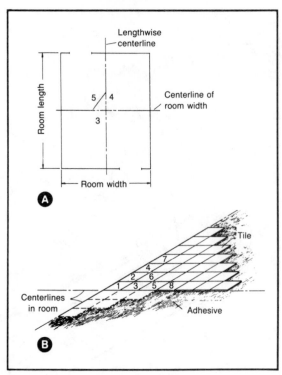

Fig. 6-9. Installing resilient tile. A: Center baseline. B: Order of laying tile. Courtesy USDA

the door width and 3 inches plus the door height. When thin resilient tile is used over the subfloor, the allowance is 2¼ inches for door height. Frames consist of a head jamb with two side jambs and the stops. When a wood casing is used around the door frame as trim, the width of the jambs is the same as the overall wall thickness. When a metal casing is used with the dry wall, eliminating the need for the wood casing, the jamb width is the same as the stud depth width. The side and head jambs and the stop are assembled as in Fig. 6-10A. Jambs may be purchased in sets or can be easily made in a small shop with a table or radial arm saw.

Casing is the trim around the door opening. It is nailed to the edge of the jamb and to the door buck (edge stud). A number of shapes are available, such as colonial (Fig. 6-10B), ranch (Fig. 6-10C), and plain (Fig. 6-10D). Casing widths vary from 2¼ to 3½ inches, depending on the style. Thicknesses vary from ½ to ¾ inches. A casing bead or metal casing used to trim the edges of the gypsum board at the

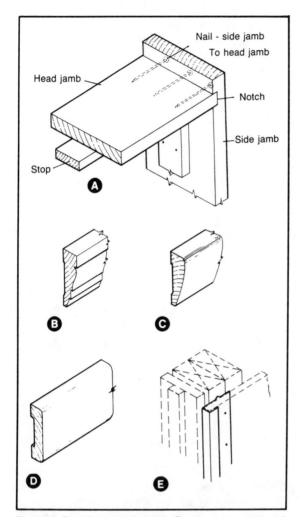

Fig. 6-10. Door frame and trim. A: Frame components and assembly. B: Colonial casing. C: Ranch casing. D: Plain casing. E: Metal casing. Courtesy USDA

ings join. Rectangular casing can be butt-joined (Fig. 6-12C). Casing for the interior side of exterior door frames is installed the same way.

If metal casing is used instead of wood casing, it can be nailed in place in two ways. After the jambs have been installed and before the gypsum board is placed, nail the metal casing to the door buck around the opening. Then insert the gypsum board into the groove and nail it to the studs in normal fashion.

The second method consists of placing the metal casing over the edge of the gypsum board, positioning the sheet properly with respect to the jambs, and nailing through both the gypsum board and the casing into the stud behind (Fig. 6-12). The same type nails and nail spacing used for gypsum board wall and ceiling finishing can be used here.

The door opening is now complete. Without the door in place, it is often referred to as a "cased opening." The door is fitted between the jambs by planing the sides to the correct width. This requires a careful measurement of the width before planing. When the correct width is obtained, the top is squared and trimmed to fit the opening. The bottom of the door is sawed off with the proper floor clearance. The side, top, and bottom clearances and the location of the hinges are shown in Fig. 6-13. The paint or finish takes up some space, so allow for this.

door and window jambs (Fig. 6-10D) eliminates the need for the wood casing.

Two general styles of interior doors are the panel and the flush door. The interior flush door is normally the hollow-core type (Fig. 6-11A). Depending on the thickness of the casing, sixpenny or sevenpenny casing or finish nails should be used. When the casing has one thin edge, use a 1½-inch brad or finish nail along this edge. Space the nails in pairs about 16 inches apart.

Casings with molded forms (Fig. 6-11B and C) must have a mitered joint where head and side cas-

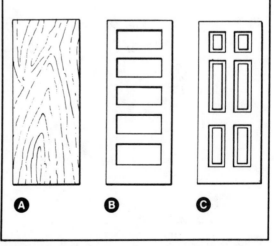

Fig. 6-11. Interior doors. A: Flush. B: five-cross-panel. C: Colonial, panel-type. Courtesy USDA

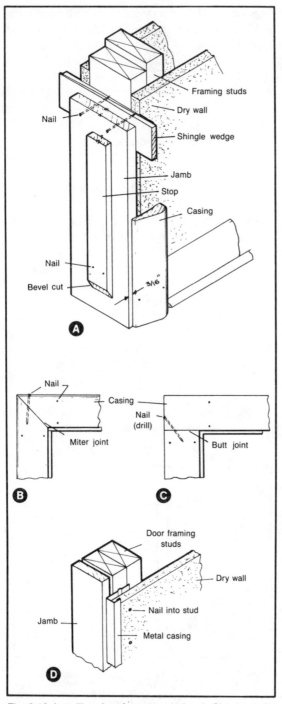

Fig. 6-12. Installing door frames and trim. A: Side jamb. B: Molded casing. C: Rectangular casing. D: Metal casing. Courtesy USDA

The narrow wood strips used as stops for the door are usually 7/16-inch thick and may be 1½ to 2¼ wide. They are installed on the jambs with a mitered joint at the junction of the head and side jambs A 45-degree bevel cut at the bottom of stops to 1½ inch above the finish floor will eliminate a dirt pocket and make cleaning or refinishing of the floor much easier.

Before fitting the exterior doors, install a threshold between side jambs to cover the junction of the sill and the flooring (or allow for the threshold). Nail to the floor and sill with finish nails.

There are four types of hardware sets commonly used for doors, and they are available in a number of finishes:

■ Entry lock for exterior doors
■ Bathroom set (inside lock control with a safety slot for opening from the outside)
■ Bedroom lock set (keyed lock)
■ Passage set (without lock).

Two hinges are normally used for 1⅜-inch interior doors. To minimize warping of exterior doors in cold climates, use three hinges. Exterior doors 1¾-inch thick require 4-by-4-inch loose-pin hinges. Interior doors 1⅜ inch thick require 3½-by-3½-inch loose-pin hinges.

Hinges are routed or mortised into the edge of the door with about a 3/16 or ¼-inch back spacing (Fig. 6-14A). This might vary slightly however. Make adjustments, if necessary, to provide sufficient edge distance so that screws have good penetration in the wood. Locate the hinges as shown in Fig. 6-13 and use one hinge half to mark the outline of the cut. If a router is not available (remember to check at rental agencies if you don't own one), mark the hinge outline and the depth of the cut, and remove the wood with a wood chisel. The depth of the routing should be such that the surface of the hinge is flush with the wood surface. Screws are included with each hinge set; use them to fasten the hinge halves in place.

Place the door in the opening and block for the proper clearances. Tack stops in place temporarily so that the door surface is flush with the edge of

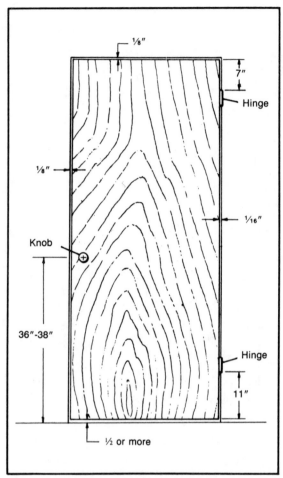

Fig. 6-13. Door clearances. Courtesy USDA

(Fig. 6-14B). Latches are used with or without a face plate, depending on the type of lock set.

The strike plate is used to hold the door closed by means of the latch. It is routed into the jamb (Fig. 6-15A). Mark the location of the latch on the jamb when the door is in a nearly closed position and outline the strike plate to this position. Route enough to bring the strike plate flush with the face of the jamb.

The stops that have been temporarily nailed in place while fitting the door and door hardware can now be permanently nailed. Exterior door frames with thicker jambs have the stop rabbeted in place as a part of the jamb. Finish nails or brads 1½ inch long are satisfactory for nailing. Nail the stop at the lock side first, setting it against the door face when the door is latched. Use nails in pairs spaced about 16 inches apart. Follow the clearances and stop locations shown in Fig. 6-15B.

Although technically considered "exterior" doors, the front and/or back door should be carefully chosen. These doors are exposed to the elements so they should be of high quality.

A solid wood door is not only functional, but the beauty of woods such as fir and hemlock make a fine first impression to the home. The rich warmth of selected woods show through clear finishes and color stains. Wood lends itself to a wide range of sculptured designs and mosaic-like effects. It's a natural

the jamb. Mark the location of the door hinges on the jambs, remove the door, and with the remaining hinge halves and a small square mark the outline. The depth of the routed area should be the same as that on the door (the thickness of the hinge half). After fastening the hinge halves, place the door in the opening and insert the pines. If marking and routing were done correctly, the door should fit perfectly and swing freely.

Door lock and latch sets are supplied with paper templates which provide the exact location of holes for the lock and latch. Follow the printed directions, locating the door knob 36 to 38inches above the floor. Most lock sets require only one hole through the face of the door and one at the edge

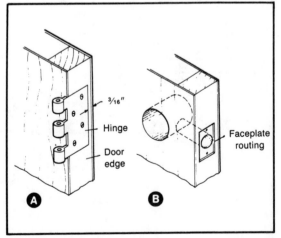

Fig. 6-14. Installing door hardware. A: Hinge. B: Lock. Courtesy USDA

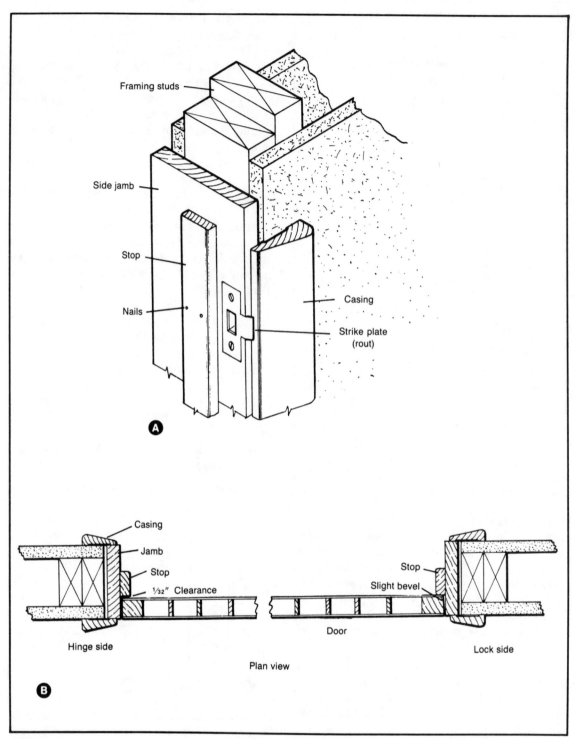

Fig. 6-15. Door installation. A: Strike plate. B: Stop clearances. Courtesy USDA

product both aesthetically pleasing and practical. It is weather-tight when properly weather-stripped, and can be easily planed to fit if settling of the home occurs. Wood is also a natural insulator.

Door Handling Guide

■ Handle all doors with clean gloves and equipment.

■ Avoid dragging doors across one another or across other surfaces.

■ Store flat in a level surface in a dry, well-ventilated building. Avoid stacking on end.

■ Cover doors to keep them clean, but allow air circulation.

■ Do not subject to abnormal heat, dryness or humidity for prolonged periods.

■ Condition all wood doors to average prevailing relative humidity of the locality before hanging.

■ Deliver doors in clean truck and under cover in wet weather.

■ Deliver doors to building site only after plaster, texture and/or cement is dry.

■ If the doors are to be stored on the job site, the entire door, including the top and bottom edges, must be sealed with a clear or pigmented base sealer in order to prevent undue moisture absorption.

Hanging Exterior Wood Doors

First, check jambs and stop; they should be square and plumb (Fig. 6-16). When trimming a door to proper height, measure the inside of the jamb from floor to header (Fig. 6-17). If replacing a door, do not go by measurements of the door you are replacing; they might not be accurate. Measure both height and width in several places because the jamb might be out of square. Allow room at bottom for sill or carpet (fig. 6-17).

Use a fine-tooth saw for trimming door to proper height across the grain of stiles (vertical wood sections). Use a wood plane for trimming to width. The best rule is to allow approximately $\frac{3}{16}$ inch clearance—under perfectly dry conditions. Weatherstripping might require more trimming, depending upon the type used, but do not trim for weatherstripping until door is hung and operational (Figs. 6-18 through 6-21).

Use three hinges on doors less than 7 feet high, four on doors over 7 feet. Position top and bottom hinges 7 to 8 inches from ends. Use a sharp wood chisel, or a special tool made for this purpose to cut insets for hinges in door stile and in door jamb (Fig. 6-22). Check frequently that hinges are installed in a straight line to prevent binding.

Hang the door on its hinges and make final trim adjustments for a precise fit before installing the lockset (Fig. 6-23). When installing lockset, follow manufacturer's instructions and use a template provided for drilling and cutting holes (Fig. 6-24). Drill from one side of the door until the point of the bit protrudes from the opposite side, then reverse the procedure and complete the drilling of the hole from the other side. Use a sharp wood chisel to cut away wood. Leave at least 1 inch of solid wood in the stile behind the lockset to maintain strength of door. Take care in installing striker plate accurately (Figs. 6-25 through 6-27).

Fitting and Hanging Tips

If re-using a lockset from a previous door, make a template of the holes in the new door. If previous hinges are used, insets need be cut only in the new door stile.

Shims (thin wood strips) can be placed behind hinges on either the sile or jamb to make small adjustments to eliminate binding or to provide a tighter closure. Final trim adjustments can be made with a plane or sand paper.

When stops are separate pieces of wood nailed to the jamb, they should be nailed into place after the door has been fitted and the lockset installed. Close the door and nail the stops securely against the door.

Check the threshold before replacing an older door in existing construction. If rotted or damaged, remove until door is hung, then replace and trim door bottom to fit. Consider replacing old thresholds with new designs that combine wood or metal plates with vinyl weatherstripping seals for greatly improved energy savings. Some thresholds are adjustable in height to accommodate uneven floors or carpet clearances.

Fig. 6-16. Jambs and stop-square and plumb. Courtesy Fir and Hemlock Assn.

Weatherstripping is important because heat and cold go around a wood door, not through it. Install weatherstripping to manufacturer's instructions, being sure to seal all edges before weatherstripping is applied.

Preparing and Finishing Doors

Wood panel doors should be finished after they have been fitted and hung. Remove the door from hinges to a dry area. Avoid finishing immediately after a rain or damp weather, or during periods of high humid-

Fig. 6-17. Measure inside of door jamb. Courtesy Fir and Hemlock Assn.

ity. The door itself must be dry.

Sand the entire surface with extra-fine sandpaper and/or steel wool to remove marks, fingerprints, stamps, etc. Grease and oil stains can be removed with mineral spirits. Do not use water or caustic or abrasive cleaners. Clean the door thoroughly with a cloth to remove dust (Figs. 6-28 and 6-29).

To obtain a quality stain or clear finish, the first step is to use an oil-base sealer manufactured for exterior use. (Do not use lacquer-base or latex-base sealers.) Sand lightly after sealer has dried (Fig. 6-30). If door is to be weatherstripped, apply two coats of sealer to edges before applying weatherstripping.

If a stain effect is desired, apply one or two coats of semi-transparent stain, letting it stand and wiping as needed to achieve the desired color. Do not

Fig. 6-18. Measuring for proper height. Courtesy Fir and Hemlock Assn.

Fig. 6-19. Use fine-tooth saw. Courtesy Fir and Hemlock Assn.

apply stain before sealing because the wood will not accept the stain evenly.

Apply at least two topcoats of an exterior oil-base clear finish. Marine-quality varnishes and polyurethane finishes give superior topcoats (Fig. 6-31).

To obtain a quality painted finish, first use a good oil-base primer manufactured for exterior use. Sand lightly when dry. The second and third coats are ''color'' coats and may be either oil-base or latex-base exterior paints.

WINDOW TRIM

The casing used around the window frames on the interior of the house is usually the same pattern as that used for the interior door frames. There are two popular methods of installing wood trim at window areas. The first is with a stool and apron (Fig. 6-32A). The second is with complete casing trim (Fig. 6-32B). Metal casing is also used around the entire window opening (Fig. 6-32C).

In a prefitted double-hung window, the stool is normally the first piece of trim to be installed. It is notched out between the jambs so that the forward edge contacts the lower sash rail. Of course, in windows that are not pre-assembled, the sash must be fitted before the trim is installed. Blind-nail the stool at the ends with eightpenny finish nails so that the casing at the side will cover the nailheads. With hardwood, predrilling is usually required to prevent splitting. The stool should also be nailed at midpoint to

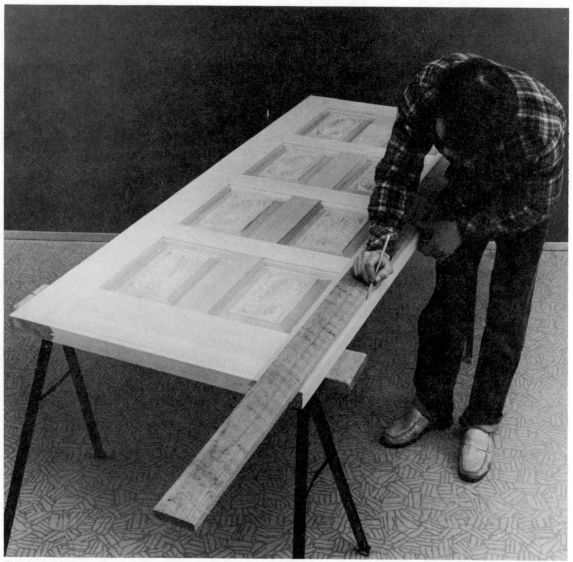

Fig. 6-20. Allow for 3/16 clearance. Courtesy Fir and Hemlock Assn.

the sill, and later to the apron when it is installed. Toenailing may be substituted for face-nailing to the sill.

Apply the casing and nail as described for the door frames, except that the inner edge is flush with the inner faces of the jambs so that the stop covers the joint. Fit the stops as the interior door stops were fitted, and place them against the lower sash so that it can slide freely. Use a 1½-inch casing nail or brad. Use nails in pairs spaced about 12 inches apart. When full-length weatherstrips are included with the window unit, locate the stops against them to provide a small amount of pressure. Cut the apron to a length equal to the outer width of the casing line, and nail to the framing sill below with eightpenny finish nails.

When casing is used to finish the bottom of the window openings, it is applied to the sill as well as

Fig. 6-21. Trim for weather stripping. Courtesy Fir and Hemlock Assn.

Fig. 6-22. Positioning hinges. Courtesy Fir and Hemlock Assn.

at the side and head of the frame. Consequently, the jambs and sill of the frame are not as deep as when wood casing is used. The stops are also narrower by the thickness of the drywall finish. The metal casing is used flush with the inside edge of the window jamb. This type of trim is installed at the same time as the drywall and in the same way described for the interior door frames.

Other types of windows, such as the awning or hopper types or the casement, are trimmed about

160

Fig. 6-23. Make final adjustments when hinges are set. Courtesy Fir and Hemlock Assn.

Fig. 6-24. Template to installing lockset. Courtesy Fir and Hemlock Assn.

Fig. 6-25. Drilling for lockset. Courtesy Fir and Hemlock Assn.

the same as the double-hung window. Casing of the same type shown in Fig. 6-19 can also be used for these units.

Some type of trim or finish is normally used at the junction of the wall and the floor. This can consist of a simple wood member that serves as both the base and the base shoe, or a more elaborate two-piece unit with a square-edge base and base cap (Fig. 6-33A). One-piece standard base may be obtained in a 2¼-inch width (Fig. 6-33B) or 3¼-inch medium width (Fig. 6-33C). The true base shoe, sometimes called quarter-round, is not actually quarter-round; it is ½ by ¾ inch in size. In the interests of economy, the base shoe can be eliminated, using only a single-piece base. Resilient floors can be finished with a simple narrow wood base or with a resilient

Fig. 6-26. Front cuts for locksets. Courtesy Fir and Hemlock Assn.

Fig. 6-27. Side cuts for lockset. Courtesy Fir and Hemlock Assn.

Fig. 6-28. Finishing door sanding. Courtesy Fir and Hemlock Assn.

cove base that is installed with adhesives.

Install wide square-edge baseboard with a butt joint at inside corners and a mitered joint at outside corners. (Fig. 6-33D). Nail it to each stud with two eightpenny finishing nails. Molded single-piece base, base molding, and base shoe should have a coped joint at inside corners and a mitered joint at outside corners. A coped joint is one in which the first piece has a square-cut end against the wall and the second member at the inside corner has a coped end. This is accomplished by sawing a 45-degree miter cut and, with a coping saw, trimming along the mi-

Fig. 6-29. Sand with steel wool. Courtesy Fir and Hemlock Assn.

Fig. 6-30. Apply sealer. Courtesy Fir and Hemlock Assn.

Fig. 6-31. Two top coats. Courtesy Fir and Hemlock Assn.

tered edge to fit the adjoining molding (Fig. 6-33E). This results in a tight inside joint. The base shoe should be nailed to the subfloor with long, slender nails, and not to the baseboard itself. This will prevent openings between the shoe and the floor if the floor joists dry out and shrink.

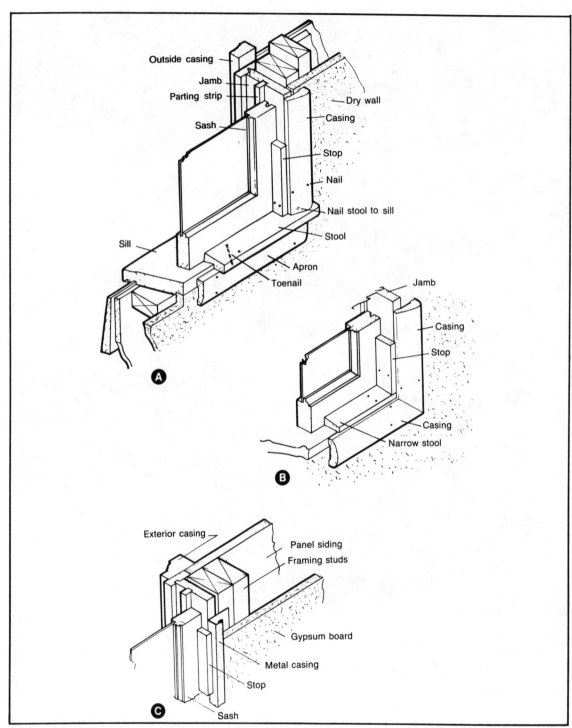

Fig. 6-32. Window trim. A: With stool and apron. B: Wood casing at window. C: Metal casing at window. Courtesy Fir and Hemlock Assn.

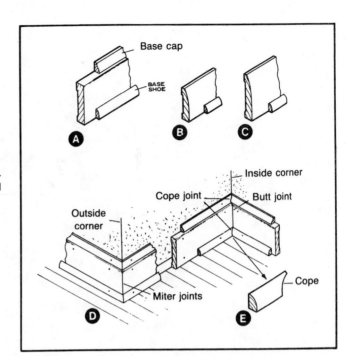

Fig. 6-33. Base moldings. A: Two-piece. B: Narrow. C: Medium width. D: Installation. E: Coped joint. Courtesy USDA

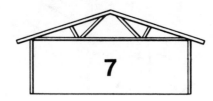

7

Auxiliary Attachments

Porches, garages, steps and even lean-to green-houses make up auxiliary attachments to a house. All may be built at the time of initial construction or added later. Additions such as a porch or an attached garage improve the appearance of a small house. In many parts of the country, the porch serves as a gathering place for family and neighbors and is considered a necessity.

The cost of porches can vary a great deal, depending on design. A fully-enclosed porch with windows and interior finish would add substantially to the cost of a house. On the other hand, the cost of an open porch with roof, floor, and simple supporting posts and footings might be well within the reach of many home builders. Such a porch could be improved in the future by enclosing all or part of it.

PORCHES

Porches could be located at either the end of a house (Fig. 7-1A) or at the side of a house (Fig. 7-1B). The porch at the end of a gable-roofed house could have a gable roof similar to the house itself. The

slope of the porch roof should usually be the same as the main roof. When the size of the porch is somewhat less than the width of the house, the roofline will be slightly lower (Fig. 7-2). When the porch is the same width, the roofline of the house is usually carried over the porch itself. A low-pitch roof in a shed or hip style can also be used for a porch located at the end of a house.

A low-slope roof can also be used for a porch at the side of a gable-roofed house (Fig. 7-3). This system of room construction is usually less expensive than other methods because the rafters also serve as ceiling joists.

A gable-roof porch at the side of the house is yet another alternative (Fig. 7-4). While somewhat more costly than a flat or low-pitch roof, it is probably more pleasing in appearance. Both rafters and ceiling joists are normally used in this system. If a porch is to be a part of the original house or constructed later, select the style most suited to the design of the house and to the available funds.

There are many types of foundations that might be used for a porch, as well as for the house. A full-

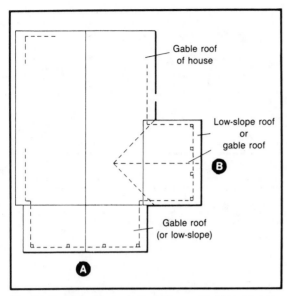

Fig. 7-1. Porch locations. A: End of house. B: Side of house. Courtesy USDA

masonry foundation wall with proper footings is substantially more expensive than the pier type. With a proper soil cover and skirtboard of some type, the performance and appearance of such a post or pier foundation will not vary greatly from that of a full-masonry wall.

Floor System

The floor system for the open porch consists of treated wood posts or masonry piers. These should bear on footings and support nail-laminated floor beams that are anchored to the posts. These beams, of doubled and nailed 2-by-8-inch or 2-by-10-inch members, serve to support the floor joists by means of ledgers (Fig. 7-5). The beams span between the wall line of the house and the posts at the outside edge of the porch.

Nail 2-by-3-inch or 2-by-4-inch ledger members to both sides of the intermediate beams and to one

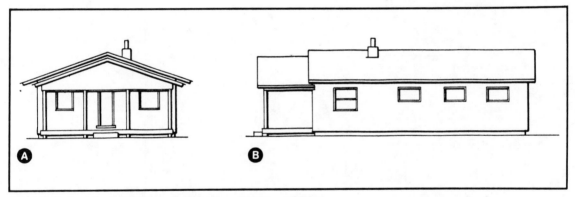

Fig. 7-2. End porch. A: End view. B: Side view. Courtesy USDA.

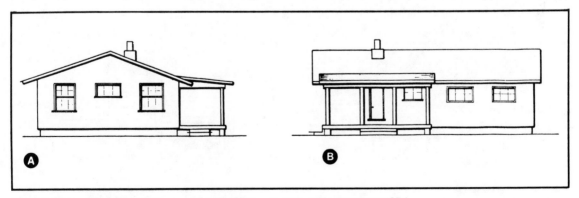

Fig. 7-3. Side porch with low-slope roof. A: End view. B: Side view. Courtesy USDA

171

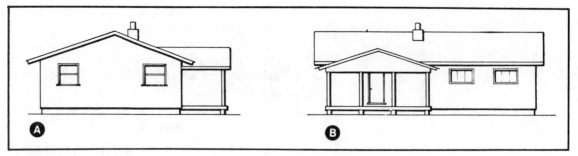

Fig. 7-4. Side porch with gable roof. A: End view. B: Side view. Courtesy USDA

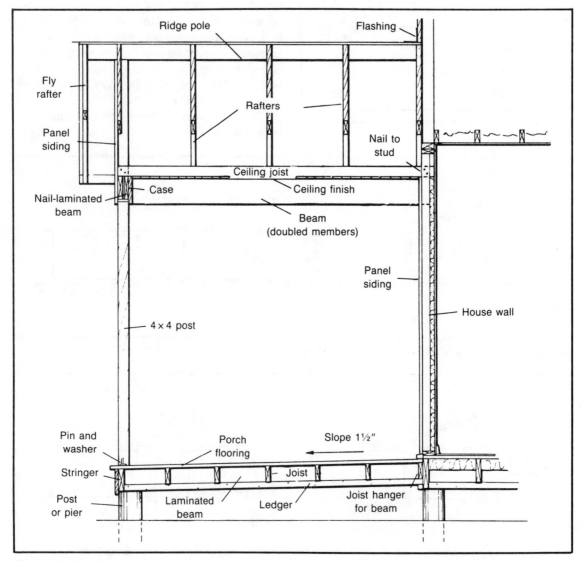

Fig. 7-5. Section through porch with gable roof. Courtesy USDA

side of single edge beams at the ends of the porch with sixteenpenny nails spaced 8 to 10 inches apart. The beams should be located so that the floor surface of an 8-foot-wide porch slopes outward a total of at least 1½ inches for drainage. Beams are spaced 10 feet or less across the width of the porch—the spacing depends on the size of joists and the beams. These details are ordinarily shown on the plans.

Joist and beam tables can also be used to determine the correct span-spacing-size relationship. Use a metal joist hanger or angle iron when fastening the beam ends to the floor framing of the house (Fig. 7-5), or allow the ends to bear on a post or pier. A single header (with ledger) the same size as the beam is used at the outside edge of the porch.

Cut the floor joists to fit between the laminated beams so that they rest on the ledgers. Space them properly according to the details in the working drawings, and toenail the ends to the beam with two eightpenny nails on each side.

Dressed and matched porch flooring, in nominal 1-by-4-inch size, can be applied to the floor joists or installed after the roof framing and roofing are in place. To protect the floor from damage, it is perhaps best to delay this phase of construction. If you do, use temporary braces for the roof until the flooring and porch posts are installed.

Framing for Gable Porch Roof

A double member or nail-laminated beam is required to carry the roof load whether a gable or low-slope roof is used. These beams are made up by using spacers of lath or plywood between 2-inch members so that the beam is the same size (3½-inch) as the nominal 4-by-4-inch solid posts used for final support of the roof. End beams made up of doubled 2-inch members are fastened to the outside beam and to the house. Thus, the outline of the porch is now formed by the front and end beams.

One method of assembling this roof framing is by nailing it together over the floor framing and raising it in place. Temporary 2-by-4-inch or larger braces are used to support this beam framing while the roof is being constructed. Use enough braces to prevent movement. Use joist hangers or a length of angle iron to fasten one end of the beam to the

house framing. The beam ends can also be carried through the wall and supported by auxiliary studs when possible. It is a good idea to provide a 6-foot, 8-inch or a 7-foot clearance from the porch floor to the bottom of the beams so that standard units can be used if screening or enclosing is added in the future.

Fasten ceiling joists to the beam at one end and to the studs of the house at the other. They should be spaced 16 or 24 inches on center to provide nailing for a ceiling material.

Rafters, either 2-by-4-inch or 2-by-6-inch, depending on the span, are now measured and cut at the ridge and wall line as described in the section on pitched roofs. Use a 1-by-6-inch ridgepole to tie each pair of rafters together. Cut gable end studs to fit between the end rafters and the outside beam. Space them to accommodate the panel siding or other finish. Toenail the studs to the outside beam and to the end rafters.

Apply the roof sheathing, the fly rafters, and the roofing as described in the sections on roof systems and coverings. Use flashing at the junction of the roof and the end wall of the house when applying the roofing.

The matched flooring can now be applied to the floor across the joists. Extend it beyond the outer edge. Use sevenpenny or eightpenny flooring nails and blind-nail to each joist. It is good practice to apply a saturating coat of water-repellent preservative on the surface and edges to provide protection until the floor is painted. Some decay-resistant wood species require little protection other than this treatment.

The nominal 4-by-4-inch posts can now be installed. When they are cut to length, drill and drive a small ⅜- or ½-inch-diameter pin into the center of one end. Drill a matching hole into the floor, apply a mastic caulk to the area, and position the post. Use a large galvanized washer between the post and the floor so moisture can evaporate. Use toenailing and metal strapping to tie the post to the roof framing. In areas of high winds, use a bolt or lag screw instead of the pin from the post to some part of the floor framing.

Matched boards, plywood, or similar covering

materials can be used to finish the ceiling surface. The exterior and interior of the laminated beams are normally eased with nominal 1-inch boards, except where siding at the gable end can be carried over the exposed face of the beam. A 1-by-4-inch member is normally used under the beam and between the posts. All exposed nailing should be done with galvanized or other rust-resistant nails.

Framing for Low-Slope Porch Roof

Framing a flat or low-slope roof for a porch is relatively simple. A nailed beam, similar to the type used for the gable roof, is required to support the ends of rafter-joists (Fig. 7-6). Single end members, however, are used to tie the beam to wall framing of the house rather than a doubled end beam. Temporary posts or braces are used to hold the beam framing after it is in place. A minimum roof slope of 1 inch per foot is desirable for drainage.

Cut the rafters and toenail them to the beam with eightpenny nails and face-nail the opposite ends to the rafter or the ceiling joists of the house. When spacing is not the same and members do not join the rafters or ceiling joists of the house, toenail the ends of the porch rafters to the wall plate. Plywood or board sheathing is applied, and when end overhang is desired, a fly rafter is added. In this case, 1-by-6-inch facia boards are applied to the ends of the rafters and to the fly rafter with sevenpenny or eightpenny galvanized siding nails.

Apply roofing as has been recommended for low-slope roofs; underlayment followed by double-leverage surfaced roll roofing. On low slopes, no exposed nails are used. A ribbon of asphalt roof cement or lap seal material is used under the lapped edge. Extend the roofing beyond the facia enough to form a natural drip edge. Ceiling covering and other trim are used in the same manner as for the gable roof.

Up to this point we have concentrated on inexpensive means of constructing porches in new wood home construction. There are, however, many excellent means of porch design that may be used during initial construction or added at a later date.

Using prehung doors, for example, is an excellent way to enclose a porch. They can dress up any area with very little installation cost and/or work (Fig. 7-7).

GREENHOUSES

There are many sizes and types of greenhouses on the market today and hundreds that come as easily constructed prefabricated packages. If you don't plan to construct your greenhouse at the time of initial construction, be sure to plan ahead for the opening. This way you can utilize glass sliding doors to what eventually will be the opening for your greenhouse/solarium.

Probably one of the largest manufacturers of greenhouses in the United States is Lord and Burn-

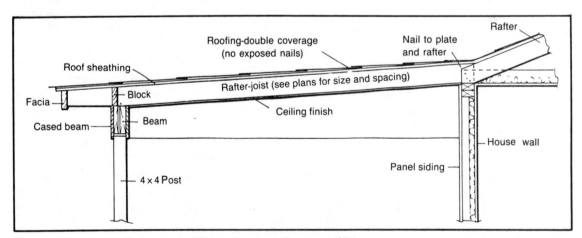

Fig. 7-6. Section through side porch with low-slope roof. Courtesy USDA

174

Fig. 7-7. Enclosed porch. Courtesy Mr. R. Mortin Miller and Hunter Morin, General Products Co. (Benchmark Doors)

ham, Irvington, New York. They have many sizes and styles from which to choose. One of the most popular is the 4-by-8 or 7-by-12 lean-to (Fig. 7-8), which fits most homes perfectly.

Foundation

When building the foundation, if there is a door in the greenhouse wall, leave about a 4-inch drop in the footing for the door buck spreader. If this was not done on your building, rest bucks on the foundation. Determine if bucks can be attached at appropriate rafter holes without raising the rafter. If not, cut the backs at top for proper connection. Cut back the glass over the door accordingly.

The normal wall height can be raised or lowered to suit local conditions. If the wall is raised, increase the height of the door sill an equal amount, and build a sloping ramp or steps as needed. Or, order long door buck extrusions and fabricate to fit during construction. If the wall height is lowered, build an excavated areaway for the door or raise the door and

frame and cut the glass over the door to fit. The door can be raised as much as 9 inches for lean-to's. Then remove soil inside the greenhouse to provide headroom.

For a greenhouse attached to another building (lean-to), determine the furthest projection on the building face in the area where the greenhouse is connected, and drop a plumb-bob to establish a starting point for construction of masonry footing.

Your greenhouse may be erected on hollow concrete or cinderblocks, poured concrete, or brick foundation. The foundation should always rest on solid ground, with footings to below frost level, approximately 36 inches deep. This is the required depth in most areas, but it will vary depending upon your location.

For supplemental greenhouse ventilation you can readily install a standard basement sash in the masonry wall. Sash of this type in steel, wood, or aluminum is available through your local building supply dealers. In warm weather, the wall sash and roof

175

Fig. 7-8. Lean-to greenhouse. Courtesy Lord and Burnham

vents can be opened and you will have steady extra ventilation because of the flue effect.

Figure 7-9 shows type foundation to be laid for a typical lean-to greenhouse and the footings that will be needed.

Installation of Sills: Accurate installation of the greenhouse sills is important. Place sill on the wall, using wood block about 1 inch thick, plus shingle wedges for accurate leveling. Nail through 1-inch blocks and on each side of sill (Fig. 7-10) to prevent movement of sills while greenhouse frame is being assembled and glass installed. Make sure aluminum sill is straight and level, with all corners at right angles. *Do not* cement sills until greenhouse is glazed; the nails through the 1-inch wood blocks will hold the frame in place and allow for any required adjustment while glazing. Verify that the width between sills is the same at the front, center, and back of the greenhouse.

Cementing the Sills: Cementing the sills should be done after glazing. Glaze the greenhouse before cementing sills because the glass will square up the framework.

Cement in the sill anchors: for a hollow block wall stuff newspapers below anchors to avoid excess use of cement.

Before cementing sills, check again to make sure they are square and level. Stretch a taut cord along the sills from end to end to make sure they are straight.

Remove wood leveling blocks and cement in sills. Work the grout well up under the sills; do not finish with a feather edge, but leave a well-rounded thickness of cement to prevent chipping. Do not work the exposed grout higher than the bottom edge of the sill both front and back.

A slate or stone cap can be used instead of concrete grout (Fig. 7-11). Allow sill anchors to set for 24 hours before removing leveling blocks and proceeding with grouting.

Framing Against Building for Lean-To

Cut a 2-by-6-inch joist 8 feet, 5¾ inches for a 4-section lean-to, 12 feet, 7¼ inches for a 6-section lean-to. Fasten joist to building so the top is 8 feet, 2 inches above location line of the greenhouse alu-

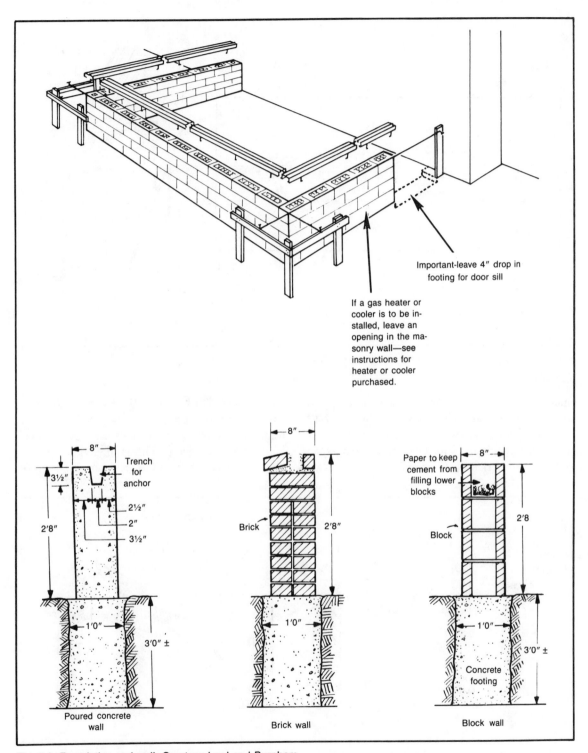

Important-leave 4″ drop in footing for door sill

If a gas heater or cooler is to be installed, leave an opening in the masonry wall—see instructions for heater or cooler purchased.

Poured concrete wall

8″

3½″

2′8″

Trench for anchor

2½″
2″
3½″

1′0″

3′0″ ±

Brick wall

8″

Brick

2′8″

1′0″

3′0″ ±

Block wall

Paper to keep cement from filling lower blocks

8″

2′8

Block

Concrete footing

1′0″

3′0″ ±

Fig. 7-9. Foundation and wall. Courtesy Lord and Burnham

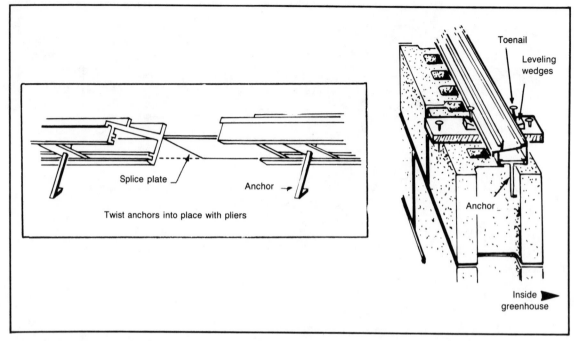

Fig. 7-10. Installing the sills. Courtesy Lord and Burnham

minum sill. Level and shim joist, and place in position on top of the 2-by-4 as shown in Fig. 7-12. Fasten with lag screws into wood clapboard surfaces; use toggle bolts for block walls, and expansion bolts on brick or stone walls.

General Framework Assembly

Sills: After the sills are set on 1-by-2-by-11-inch temporary wood blocks, make sure they are absolutely level and square. This is the most important step in the assembly of the greenhouse structure.

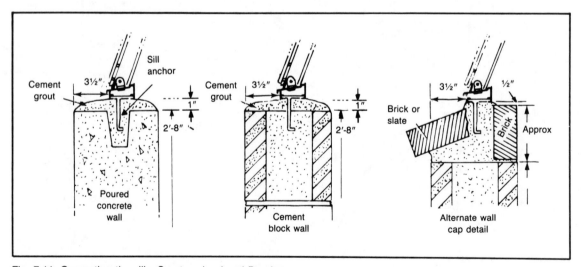

Fig. 7-11. Cementing the sills. Courtesy Lord and Burnham

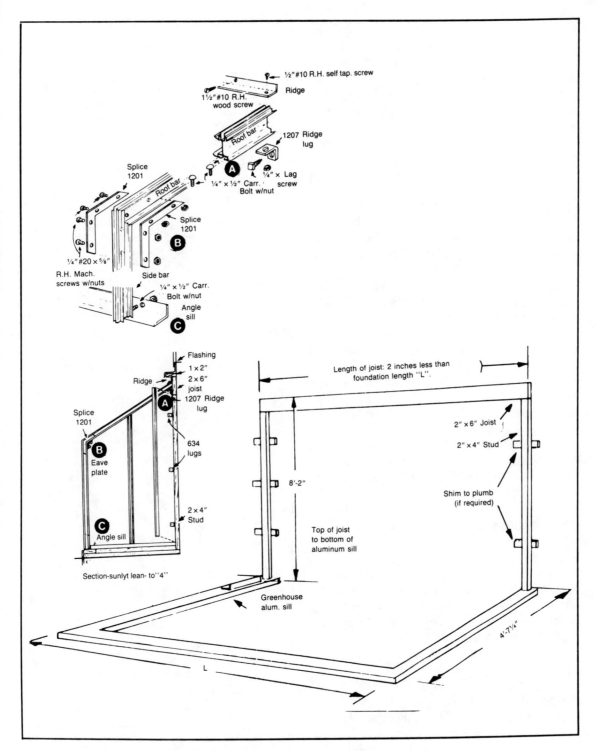

½" #10 R.H. self tap. screw

1½" #10 R.H. wood screw

Ridge

Roof bar

1207 Ridge lug

A

¼" × ½" Carr. Bolt w/nut

¼" × Lag screw

Splice 1201

Roof bar

Splice 1201

B

¼" #20 × ⅝" R.H. Mach. screws w/nuts

Side bar

¼" × ½" Carr. Bolt w/nut

Angle sill

C

Flashing

1 × 2"

2 × 6" joist

1207 Ridge lug

Ridge

A

Splice 1201

634 lugs

B

Eave plate

2 × 4" Stud

C

Angle sill

Section-sunlyt lean-to "4"

Length of joist: 2 inches less than foundation length "L".

2" × 6" Joist

2" × 4" Stud

8'-2"

Shim to plumb (if required)

Top of joist to bottom of aluminum sill

Greenhouse alum. sill

4'-7¼"

L

Fig. 7-12. Framing against building for lean-to. Courtesy Lord and Burnham

Take several measurements down the length of the greenhouse to verify that the dimensions of the inside face of the sill are correct, and exactly the same at each location checked. Also check diagonally from corner to corner. These dimensions should be the same. Make sure spacing between gable and/or side door sills is accurate (Fig. 7-13A).

Stretch a taut string down the length of each sill, and verify that it is level and straight. Do this three times: after the sills are replaced and toenailed in position on the wall; after the structure is erected; after glazing, before cementing the sill anchors.

Roof Bars, Side Bars, and Eaves: Install the roof bars, side bars, and eaves first. Do *not* erect the end rafter(s) until after vent shafting is installed (Fig. 7-13B).

Use gauges to locate the eave height. Set each gauge into the slot in the bar tongue, with the bottom edge resting on the glass seat of the sill. A pair of gauges is furnished, so that with two people working, both ends of an eave piece can be located simultaneously.

Ridge and Vent Header: Install the ridge next, then the vent header(s). Slide the vinyl plastic into the groove on the bottom of the vent header before installing the header on the roof. Snip out excess plastic where the header crosses the roof bar. Attach glazing clips to the bottom of the header.

The position of the vent header on the roof from the ridge is determined by use of gauges. Hook the end with a diagonal cut over the leg on the bottom of the ridge, and position the upstanding leg of the vent header in the notch on the lower end of the gauge.

Vent Shafting: Slide the vent shafting through the holes in the roof bars, installing and locating the arms and rods in accordance with the assembly drawing. Note also where the manual gear is located, depending on length of greenhouse, and install gear at correct location. Then install the gable end rafter(s) (Fig. 7-13C and D).

Roof Vent Sash: Assemble the roof sash on the ground in two- or three-section lengths. Do not attach the muntins at splice locations at this time. Assemble sash top rail so slots match pins in ridge. Lift the sash assembly into position on the green-

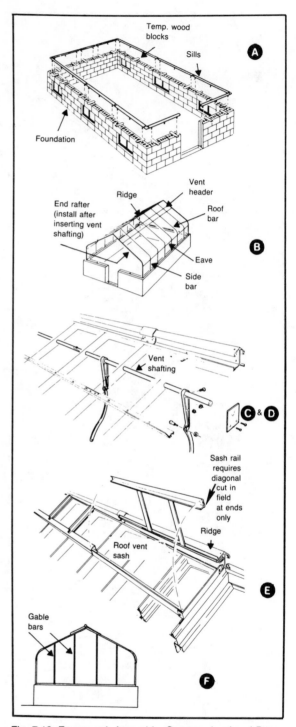

Fig. 7-13. Framework Assembly. Courtesy Lord and Burnham

house roof, and hook the sash top rail into position on the ridge in two- or three-section lengths. Then splice the sash sections together, and attach the muntins at splice locations (Fig. 7-13E).

Gable Bars: Install the gable bars, and angle purlin. If there is a door, be sure to attach the door closure channel to each door buck before placing in position (Fig. 7-13F).

After the greenhouse is fully glazed and the sills ready to be grouted, move the door closures snugly against the masonry. Attach it to the wall with angle lugs.

Sill

The sill is a horizontal member resting on top of and secured to a masonry wall. Glazing bars, end rafters, and door frame are attached to this sill. Set sill level and square, and follow dimensions shown in plans. Fittings are as follows:

■ Sill lugs—These connect glazing bars to sill.

■ Sill anchors—These twist-fit into underside of sill, to anchor sill masonry wall to concrete after glazing.

■ Sill Splice—This slides into underside of sill, where separate sill lengths come together.

■ Sill Corner—This fits into underside of wall where sills join at outer corners, and attaches to gable end rafter.

■ Door Frame Casting—This attaches to sill at door opening. Figure 7-14 provides a detailed breakdown of the sill.

Greenhouse Fittings

Web splices occur in the greenhouse at angular intersections of glazing bars (bar to bar) and end rafters (rafter to rafter). They connect these members at the eave, ridge, and other points where an angular joint is required. These points are indicated by location number 1 through 7 in Fig. 7-15.

Web splice plates are used in pairs—one on each side of the intersecting bars—except at ends, where only one splice is used on the inner side of the bar.

Glazing Bars

Glazing Bars extend from sill to ridge, and support the glass, ridge, and vent sash. They are furnished in one piece for curved eave models. Fittings are as follows:

■ Sill lugs—These connect glazing bars to side sill.

■ Web Splices—These join roof and side glazing bars at the eave line, also at the ridge on even span houses, and at the roof and deck bars of 2-bench lean-to's. Use 2 pier joints and slide into position against bar web.

■ Barcaps—These fit over and are screwed to the glazing bars when the glass is installed.

■ Bar Lug—For lean-to's, connect end of bar to joist.

Eave

The eave is a horizontal member at the intersection of roof and side, running the full length of the greenhouse. It receives top of side glass, and supports bottom of roof glass. Use a gauge to position height of the eave from the sill (Fig. 7-16). Fittings are as follows:

■ Eave Splice—This is the connecting piece where separate lengths of eave come together.

■ End Clip—This fastens to the outer ends of the eave and ridge for attaching the eave and ridge to the end rafter of the bar (Fig. 7-17).

Vent Header

The vent header is a longitudinal member approximately 26 inches below the greenhouse ridge which forms the bottom closure for the roof sash. To locate, use gauges furnished with your prefabricated greenhouse. Fittings are as follows:

■ Glazing Clips—These support the upper edge of roof glass at the header, in the center between bars. Attach to underside of header before glazing.

■ Plastic Weather Seal—Insert these in underside of header and position for bar spacing. They must be installed before header is attached to roof bars. Use soap and water as a lubricant.

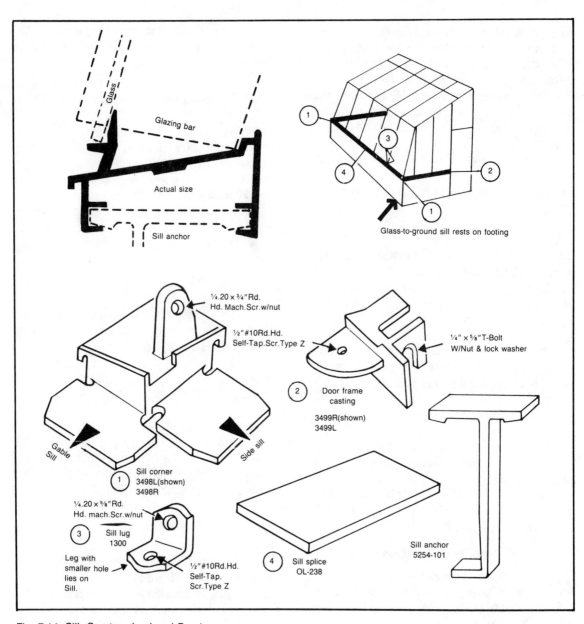

Fig. 7-14. Sill. Courtesy Lord and Burnham

■ Header Splice—This is the connecting piece where separate lengths of header come together (Fig. 7-18).

Positioning the Eave and Vent Header

To locate the eave height, use gauges. Set each gauge into the slot in the bar tongue, with the bottom edge resting on the glass seat of the sill. A pair of gauges is furnished, so that with two people working both ends of an eave piece can be located simultaneously (Fig. 7-19).

The position of the vent header down the roof

182

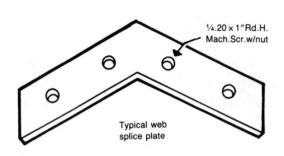

¼.20 × 1″Rd.H.
Mach.Scr.w/nut

Typical web
splice plate

Greenhouse Model No.	Location numbers						
	1	2	3	4	5	6	7′/7a
0-10	1006	1004	1006			1004	
0-14	1006	1004	1368			1005	1366
C-14			1368			1005	1366
I-14			1006			1004	
0-7	1006	1004					1366
C-7							1366
I-⅞				1314	1325		
0-⅞	1006	1004		1345	1346		1366
C-⅞				1345	1346		1366
Splice plate numbers							

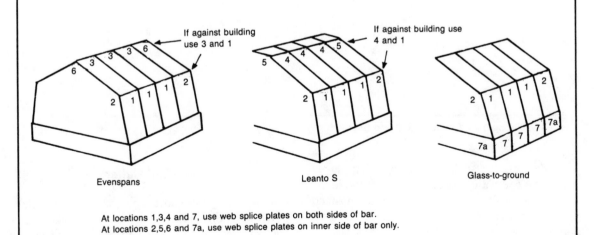

If against building use 3 and 1

If against building use 4 and 1

Evenspans

Leanto S

Glass-to-ground

At locations 1,3,4 and 7, use web splice plates on both sides of bar.
At locations 2,5,6 and 7a, use web splice plates on inner side of bar only.

Fig. 7-15. Web splices. Courtesy Lord and Burnham

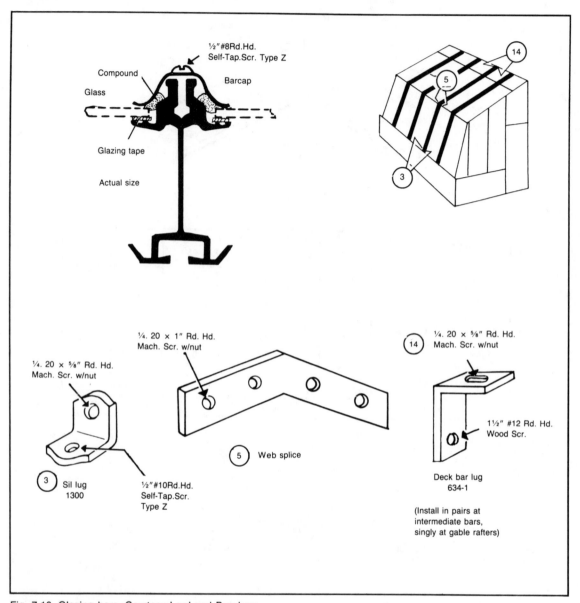

Fig. 7-16. Glazing bars. Courtesy Lord and Burnham

from the ridge is also determined by using gauges. Hook the end with a diagonal cut over the leg on the bottom of the ridge, and position the upstanding leg of the vent header in the notch on the lower end of the gauge (Fig. 7-20).

Figure 7-21 provides a complete breakdown of the ridge section to this lean-to greenhouse.

Gable-End Rafters

This section applies only to houses with gable end(s) as shown in Fig. 7-22. When house is connected to another building, a roof bar is provided, plus special ridge end closure(s), lugs to connect locally purchased wood scribing for the end bar, and short

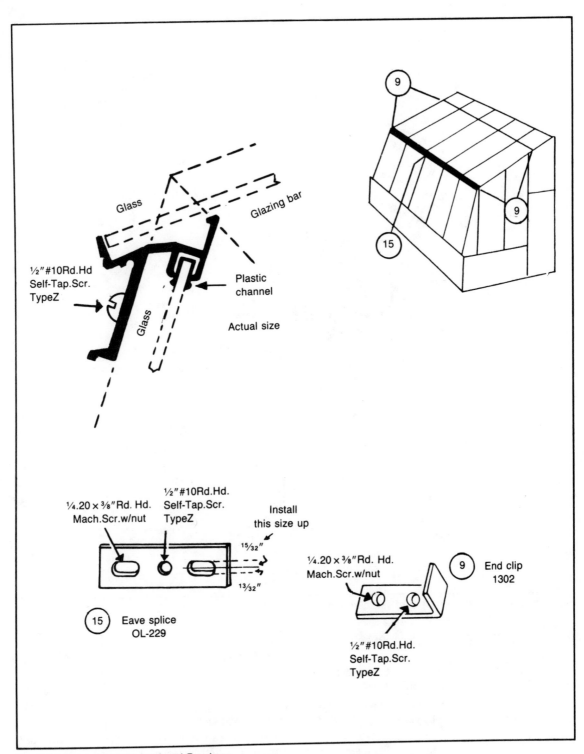

Glass

Glazing bar

½″#10Rd.Hd
Self-Tap.Scr.
TypeZ

Plastic
channel

Glass

Actual size

9

15

9

¼.20 × ⅜″Rd. Hd.
Mach.Scr.w/nut

½″#10Rd.Hd.
Self-Tap.Scr.
TypeZ

Install
this size up

15/32″

13/32″

15 Eave splice
OL-229

¼.20 × ⅜″Rd. Hd.
Mach.Scr.w/nut

9   End clip
1302

½″#10Rd.Hd.
Self-Tap.Scr.
TypeZ

Fig. 7-17. Eave. Courtesy Lord and Burnham

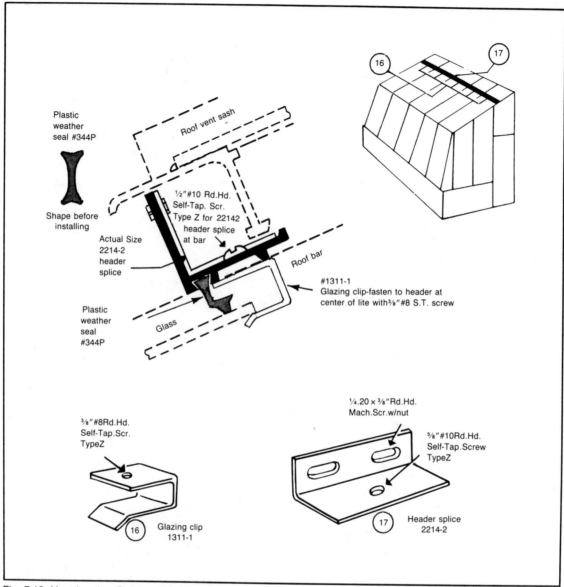

Fig. 7-18. Vent header. Courtesy Lord and Burnham

length(s) of sill, which might have to be cut to fit space.

End rafters extend from sill to ridge at the end(s) of a greenhouse with a gable. This is furnished by the prefabricator in one piece for curved eave models. Fittings are as follows:

■ Sill Corner Casing—This connects end rafter(s) to side sill.

■ Web Splices—These join roof and side end rafters at eave line. They also join at ridge on even span houses, and at roof and deck bars of 2-bench lean-to's. On gable-end rafters, use one splice per joint and slide into position on the inner side of the rafter. On locations other than gable end rafter, use in pairs (on both sides of member)

186

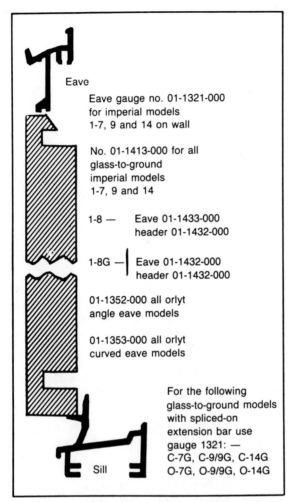

Eave

Eave gauge no. 01-1321-000
for imperial models
1-7, 9 and 14 on wall

No. 01-1413-000 for all
glass-to-ground
imperial models
1-7, 9 and 14

1-8 —   Eave 01-1433-000
          header 01-1432-000

1-8G —   Eave 01-1432-000
          header 01-1432-000

01-1352-000 all orlyt
angle eave models

01-1353-000 all orlyt
curved eave models

For the following
glass-to-ground models
with spliced-on
extension bar use
gauge 1321: —
C-7G, C-9/9G, C-14G
O-7G, O-9/9G, O-14G

Sill

Fig. 7-19. Positioning the eave and vent header. Courtesy Lord and Burnham

■ Vent Shaft Bearers—These attach to the inside face of end rafters, to support vent pipe shaft.

Gable Bars

Gable bars are vertical members in the end(s) of the greenhouse that receive and support the gable glass. Fittings are as follows:

■ T-bolts—These connect the top of vertical gable members (bars and door bucks) to end rafters.
■ Angle Purlin—This is a cross brace attached to back of bars with a ¼-by-⅞ inch T-bolt with nut and lock washer.

■ Barcaps—These fit over and are screwed to the glazing bars when the glass is installed.
■ Sill Lugs—These connect gable bars to the sill.
■ Door Frame Casting—This attaches to the sill at the door opening.
■ Door Closures—These are adjustable closures between the wall and door bucks.
■ Spreader—These attach the angle door buck spreader at the lower end (Fig. 7-23).

Vent Sash Assembly

The vent sash assembly is located next to ridge of the greenhouse. It is connected to ridge to ventilate the greenhouse by allowing warm air to flow out when the sash is open. When possible, install an automatic vent sash on side away from prevailing winter winds. The sash is a continuous assembly from end to end of the greenhouse—except in larger and partitioned houses. Fittings are as follows:

■ Top Rail—This slot in the left end engages an anti-drift pin in the ridge. The remainder of the fittings are: top rail splice, bottom rail, bottom rail splice, muntins, arms and rods, vent motors, vent shaft, nuts, bolts, glass, barcaps, sash bearers (Figs. 7-24 and 7-25).

Door and Door Framing

If you are using the type of greenhouse that attaches to your house, you have a choice of doors. You could use glass sliding doors into the main house, whereby

Table 7-1. Measurements for Existing Door Openings.

Pre-Hung Unit	To Fit Existing Openings	
	Widths (A)	Heights (B)
5'4'	67"	81¾"
6'0"	75"	81¾"
8'0"	99⅛"	81¾"
9'0"	111⅛"	81¾"

Courtesy Mr. R. Mortin Miller and Hunter Morin, General Products Co.

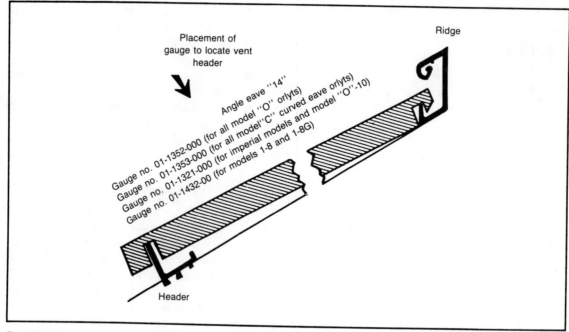

Fig. 7-20. Aluminum gauge for locating vent header. Courtesy Lord and Burnham

you may close off the greenhouse if desired. You could also use a door that opens to the outside.

Be sure to leave a 4-inch drop in the door footing for the door sill spreader, and to permit a sufficient amount of concrete when grouting to provide a solid threshold.

Gable, partition, and side doors are furnished pre-assembled with your prefabricated greenhouse model. They are framed and hinged for either the right or left hand. If you plan to use this prefabricated (or any other prefabricated model) be sure to specify whether you want left- or right-hand doors with your greenhouse order. For partition and side entrance door installations, refer to the special drawings that come with your greenhouse.

To install the door, attach the gable purlin with T-bolts and lock washers. Leave T-bolts loose for adjustment. Use ¼-inch spacer between all bars or bucks when purlin extends to gable rafter.

Attach door head bucks to the angle purlin with 5¼-by-⅞-inch T-bolts and lock washers.

At the bottom edge of the door, cut door frame 1⅛ inch (Fig. 7-26). Use a hacksaw to cut that part of the frame that conflicts with spreader. Place ⅛-inch shims on angle spreader. Position the door frame between door bucks and lower door head buck and angle purlin to rest on top of the frame. Center door and frame on door bucks. Drill ⁵⁄₃₂-inch holes into door bucks where holes appear in the door frame.

Doors have one rivet on the lock side that should be removed after the door and frame are connected to the door bucks, using a sharp chisel or hacksaw blade to cut the rivet. Open the door wide and drill ⁵⁄₃₂-inch holes at each hinge location. Drive ½-inch screws until screw head seats. Do not tighten.

Attach glazing adaptor to door head buck with three ½-inch screws. Figures 7-27, 7-28, and 7-29 provide detailed illustrations of the door installation. Figure 7-30 shows the details of caulking the greenhouse to the main building.

GARAGES

Garages can be classified as attached, detached, basement, or carport. The selection of a garage is often determined by limitations of the site and the size of the lot. Where space is not a limitation, the attached garage has much in its favor. It can give better architectural lines to the house, it is warmer

188

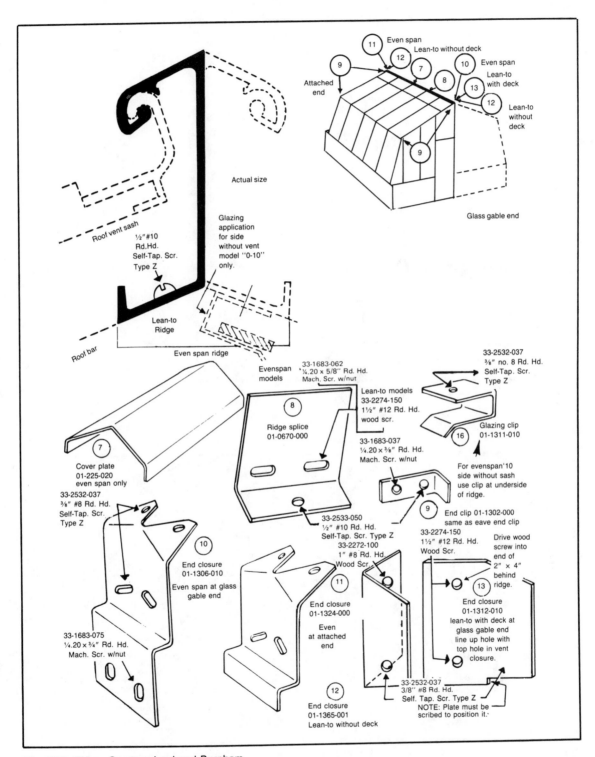

Fig. 7-21. Ridge. Courtesy Lord and Burnham

Actual size

Roof vent sash

½″ #10
Rd. Hd.
Self-Tap. Scr.
Type Z

Glazing
application
for side
without vent
model ''0-10''
only.

Lean-to
Ridge

Roof bar

Even span ridge

Even span

Lean-to without deck

Even span

Attached
end

Lean-to
with deck

Lean-to
without
deck

Glass gable end

Evenspan
models

33-1683-062
¼.20 x 5/8″ Rd. Hd.
Mach. Scr. w/nut

Lean-to models
33-2274-150
1½″ #12 Rd. Hd.
wood scr.

33-2532-037
⅜″ no. 8 Rd. Hd.
Self-Tap. Scr.
Type Z

Ridge splice
01-0670-000

33-1683-037
¼.20 x ⅜″ Rd. Hd.
Mach. Scr. w/nut

Glazing clip
01-1311-010

For evenspan'10
side without sash
use clip at underside
of ridge.

Cover plate
01-225-020
even span only

33-2532-037
⅜″ #8 Rd. Hd.
Self-Tap. Scr.
Type Z

33-2533-050
½″ #10 Rd. Hd.
Self-Tap. Scr. Type Z

33-2272-100
1″ #8 Rd. Hd.
Wood Scr.

End clip 01-1302-000
same as eave end clip

33-2274-150
1½″ #12 Rd. Hd.
Wood Scr.

Drive wood
screw into
end of
2″ x 4″
behind
ridge.

End closure
01-1306-010
Even span at glass
gable end

End closure
01-1324-000

Even
at attached
end

End closure
01-1312-010
lean-to with deck at
glass gable end
line up hole with
top hole in vent
closure.

33-1683-075
¼.20 x ¾″ Rd. Hd.
Mach. Scr. w/nut

End closure
01-1365-001
Lean-to without deck

33-2532-037
3/8″ #8 Rd. Hd.
Self. Tap. Scr. Type Z
NOTE: Plate must be
scribed to position it.

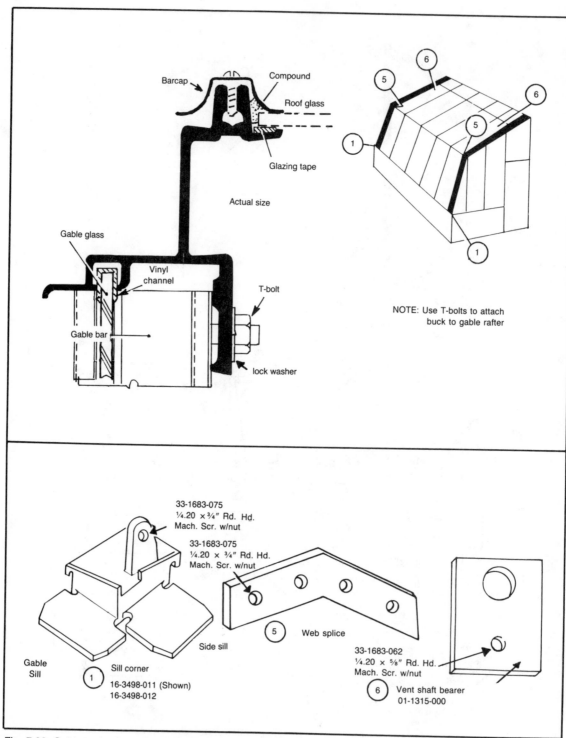

Fig. 7-22. Gable-end rafters. Courtesy Lord and Burnham

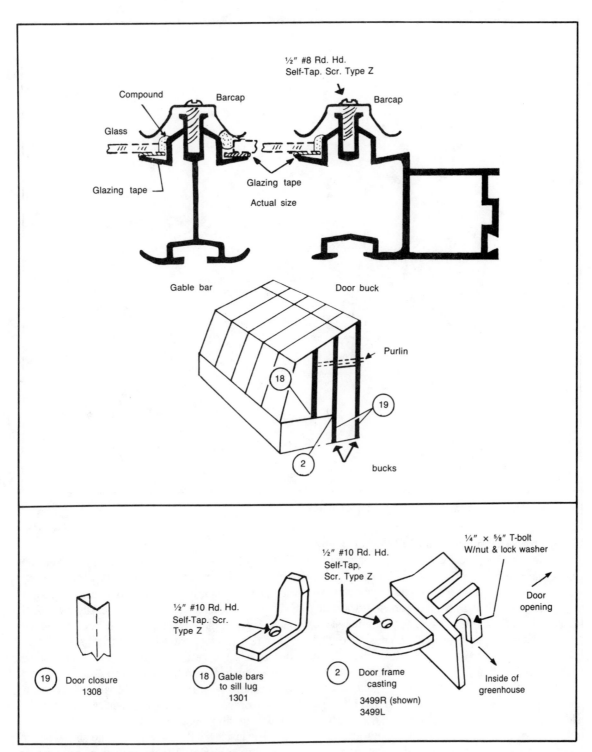

½″ #8 Rd. Hd.
Self-Tap. Scr. Type Z

Compound   Barcap   Barcap

Glass

Glazing tape

Glazing tape

Actual size

Gable bar   Door buck

Purlin

18

19

2

bucks

¼″ × ⅝″ T-bolt
W/nut & lock washer

½″ #10 Rd. Hd.
Self-Tap.
Scr. Type Z

Door
opening

½″ #10 Rd. Hd.
Self-Tap. Scr.
Type Z

Inside of
greenhouse

19 Door closure
    1308

18 Gable bars
    to sill lug
    1301

2 Door frame
   casting

   3499R (shown)
   3499L

Fig. 7-23. Gable bars. Courtesy Lord and Burnham

191

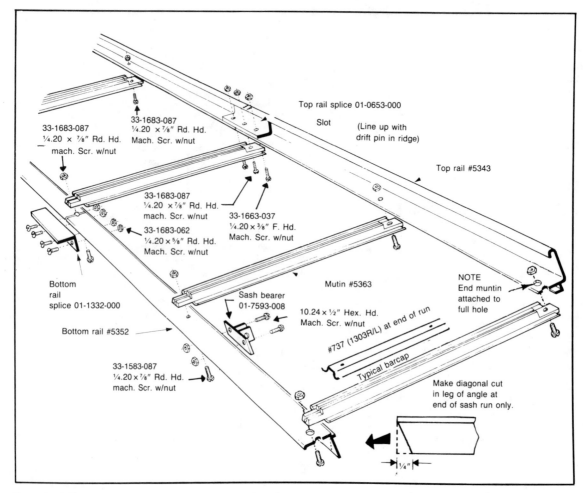

Fig. 7-24. Vent sash assembly. Courtesy Lord and Burnham

during cold weather, and it provides covered protection to passengers, convenient space for storage, and a short, direct entrance to the house.

Building regulations often require that detached garages be located away from the house toward the rear of the lot. Where there is considerable slope to a lot, basement garages might be desirable, and generally such garages will cost less than those above grade.

Carports are car storage spaces, generally attached to the house, that have roofs and often no sidewalls. To improve the appearance and utility of this type of structure, storage cabinets are often used on a side and at the end of the carport.

Size

Don't make the mistake of designing a garage too small for convenient use. Cars vary in size. Many popular models are 215 inches long, and the larger and more expensive models are easily over 230 inches. Even the new smaller cars take up enough room that you will lose a lot of storage space if you don't build the garage large enough. It is wise to provide a minimum distance of 21 to 22 feet between the inside face of the front and rear walls. If additional storage or work space is required at the back, allow a greater depth.

The inside width of a single garage should never be less than 11 feet; 13 feet is much more satisfac-

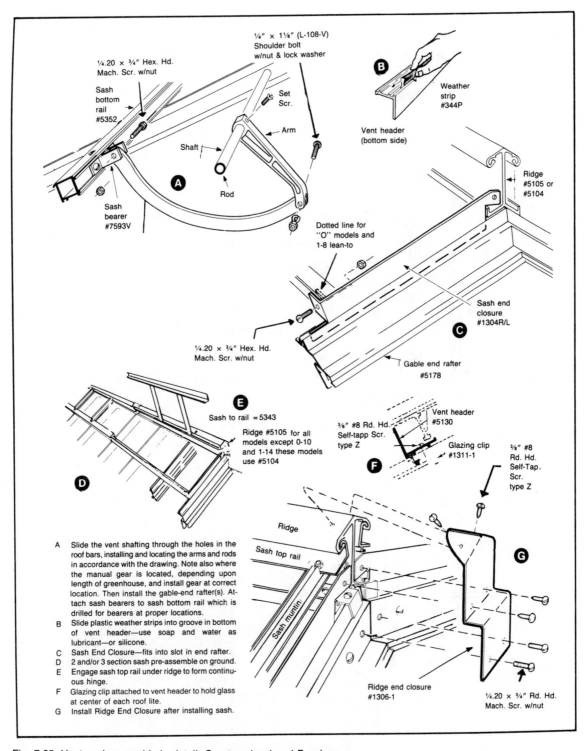

Fig. 7-25. Vent sash assembly in detail. Courtesy Lord and Burnham

Labels within figure:

¼.20 × ¾" Hex. Hd. Mach. Scr. w/nut

Sash bottom rail #5352

Sash bearer #7593V

Shaft

Arm

Rod

Set Scr.

¼" × 1⅛" (L-108-V) Shoulder bolt w/nut & lock washer

Weather strip #344P

Vent header (bottom side)

Ridge #5105 or #5104

Dotted line for "O" models and 1-8 lean-to

¼.20 × ¾" Hex. Hd. Mach. Scr. w/nut

Sash end closure #1304R/L

Gable end rafter #5178

Sash to rail = 5343

Ridge #5105 for all models except 0-10 and 1-14 these models use #5104

⅜" #8 Rd. Hd. Self-tapp Scr. type Z

Vent header #5130

Glazing clip #1311-1

⅜" #8 Rd. Hd. Self-Tap. Scr. type Z

Ridge

Sash top rail

Sash muntin

Ridge end closure #1306-1

¼.20 × ¾" Rd. Hd. Mach. Scr. w/nut

A   Slide the vent shafting through the holes in the roof bars, installing and locating the arms and rods in accordance with the drawing. Note also where the manual gear is located, depending upon length of greenhouse, and install gear at correct location. Then install the gable-end rafter(s). Attach sash bearers to sash bottom rail which is drilled for bearers at proper locations.

B   Slide plastic weather strips into groove in bottom of vent header—use soap and water as lubricant—or silicone.

C   Sash End Closure—fits into slot in end rafter.

D   2 and/or 3 section sash pre-assemble on ground.

E   Engage sash top rail under ridge to form continuous hinge.

F   Glazing clip attached to vent header to hold glass at center of each roof lite.

G   Install Ridge End Closure after installing sash.

193

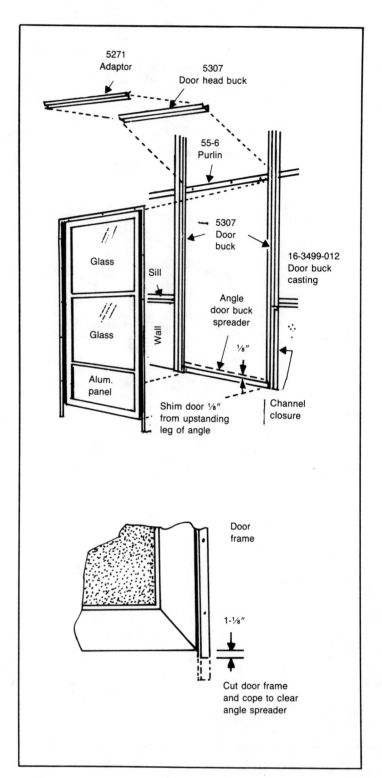

Fig. 7-26. Installing the door. Courtesy Lord and Burnham

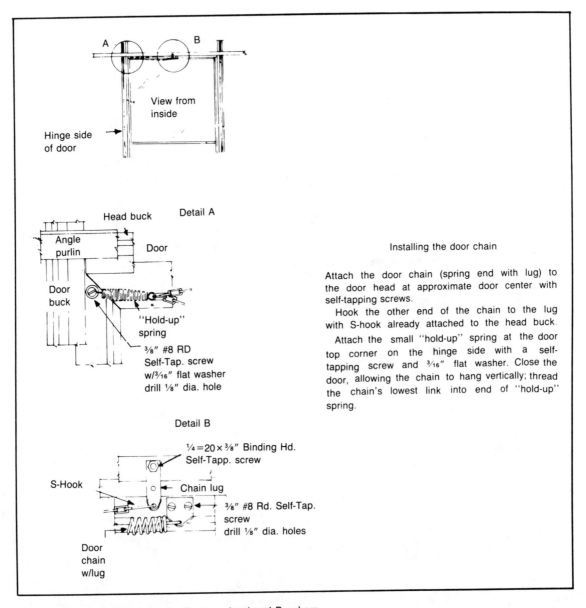

A   B

View from
inside

Hinge side
of door

Detail A

Head buck

Angle
purlin

Door

Door
buck

"Hold-up"
spring

⅜" #8 RD
Self-Tap. screw
w/³⁄₁₆" flat washer
drill ⅛" dia. hole

Installing the door chain

Attach the door chain (spring end with lug) to the door head at approximate door center with self-tapping screws.

Hook the other end of the chain to the lug with S-hook already attached to the head buck.

Attach the small "hold-up" spring at the door top corner on the hinge side with a self-tapping screw and ³⁄₁₆" flat washer. Close the door, allowing the chain to hang vertically; thread the chain's lowest link into end of "hold-up" spring.

Detail B

¼ =20 × ⅜" Binding Hd.
Self-Tapp. screw

S-Hook

Chain lug

⅜" #8 Rd. Self-Tap.
screw
drill ⅛" dia. holes

Door
chain
w/lug

Fig. 7-27. Door installation details. Courtesy Lord and Burnham

tory. The minimum outside size for a single garage, therefore, would be 14 feet by 22 feet. A double garage should not be less than 22 feet by 22 feet in outside dimensions to provide reasonable clearance and use.

For an attached garage, the foundation wall should extend below the frostline and about 8 inches above the finished floor level. It should be not less than 6 inches thick, but is usually more because of the difficulty of trenching this width. The sill plate should be anchored to the foundation wall with anchor bolts spaced about 8 feet apart, at least two bolts in each sill piece. Extra anchors might be required at the side of the main door. The framing of

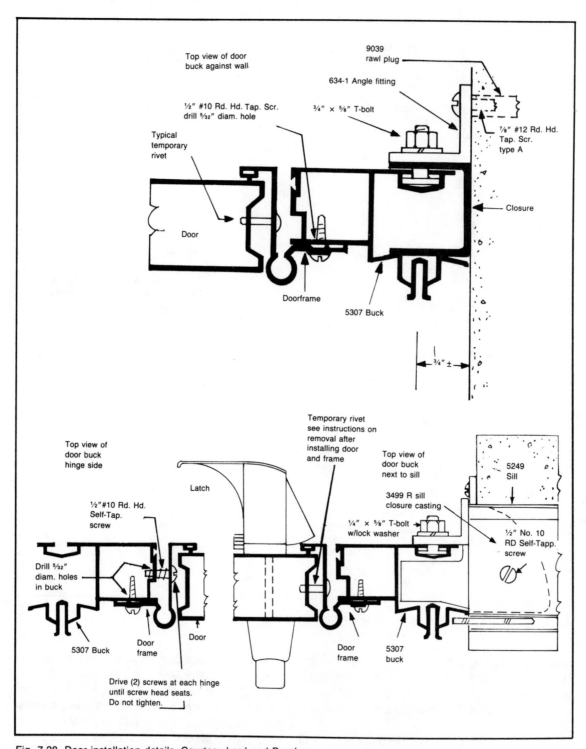

Fig. 7-28. Door installation details. Courtesy Lord and Burnham

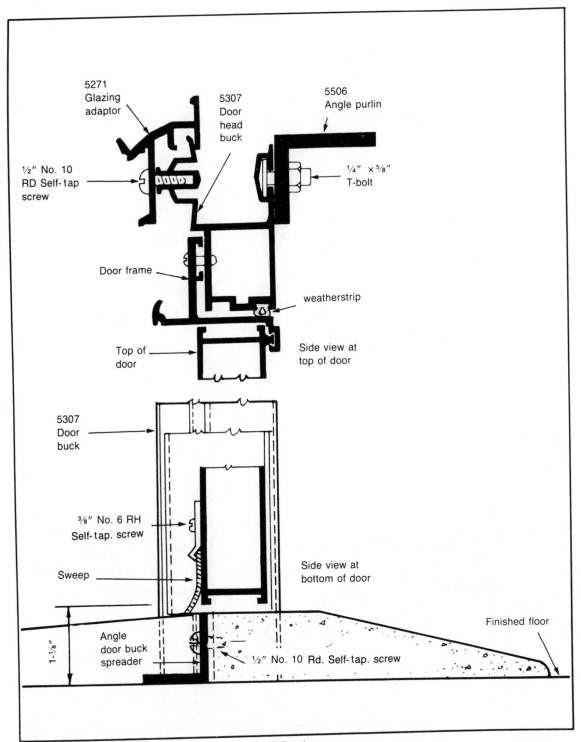

Fig. 7-29. Door installation details. Courtesy Lord and Burnham

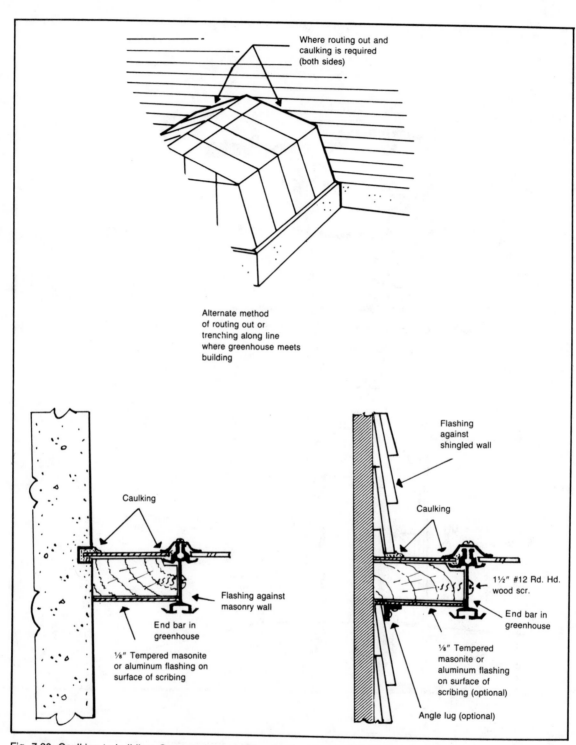

Where routing out and
caulking is required
(both sides)

Alternate method
of routing out or
trenching along line
where greenhouse meets
building

Flashing
against
shingled wall

Caulking

Caulking

Flashing against
masonry wall

End bar in
greenhouse

1½" #12 Rd. Hd.
wood scr.

End bar in
greenhouse

⅛" Tempered masonite
or aluminum flashing on
surface of scribing

⅛" Tempered
masonite or
aluminum flashing
on surface of
scribing (optional)

Angle lug (optional)

Fig. 7-30. Caulking to building. Courtesy Lord and Burnham

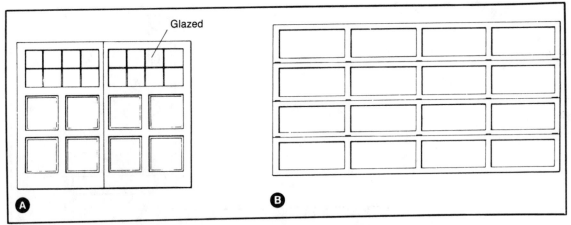

Fig. 7-31. Garage doors. A: One-section swing. B: Sectional. Courtesy USDA

the sidewalls and roof and the application of the exterior covering material of an attached garage should be similar to that of the house.

The interior finish of the garage is usually the builder's choice. The studs may be left exposed or covered with some type of sheet material, or they may be plastered. Some building codes require that the wall between the house and the attached garage be made of fire-resistant material. Consult local building regulations and fire codes before beginning construction.

If fill is required below the floor, it should be sand or gravel that is well-compacted and tamped. If other types of soil fill are used, wet it down so that it will be well compacted. Then tamp it and allow enough before pouring. Unless these precautions are taken, the concrete floor could settle and crack.

The floor should be of concrete not less than 4 inches thick and laid with a pitch of about 2 inches from the back to the front of the garage. It is advisable to use wire reinforcing mesh. The garage floor should be set about 1 inch above the drive or apron level. At this point it is good to have an expansion joint between the garage floor and the driveway or apron.

Garage Doors

The two most popular overhead garage doors are the sectional and the single-section swing types. The swing door (Fig. 7-31A) is hung with side and overhead brackets and an overhead track, and must be moved outward slightly at the bottom as it is opened. The sectional type (Fig. 7-31B) in four or five horizontal hinged sections, has a similar track extending along the sides and under the ceiling framing, with a roller for the side of each section. It is opened by lifting and is adaptable to remote-control opening devices. The standard size for a single door is 9 feet wide by 6½ or 7 feet high. Double doors are usually 16 by 16½ feet or 7 feet in size.

Doors vary in design, but those most often used are the panel type with solid stiles and rails and panel fillers. A glazed panel section is often included. The required clearance above overhead doors is usually about 12 inches. However, low headroom brackets are available when such clearance is not possible.

The header beam over garage doors should be designed for the snow load which might be imposed on the roof above. In wide openings, this can be a steel I-beam or a built-up wood section. For spans of 8 or 9 feet, two doubled 2-by-10s of high-grade Douglas fir or similar species are commonly used when only snow loads must be considered. If floor loads are also imposed on the header, a steel I-beam or wide-flange beam is often used.

STEPS AND STAIRS

Outside entry platforms and steps are required for most types of wood-frame houses. Houses con-

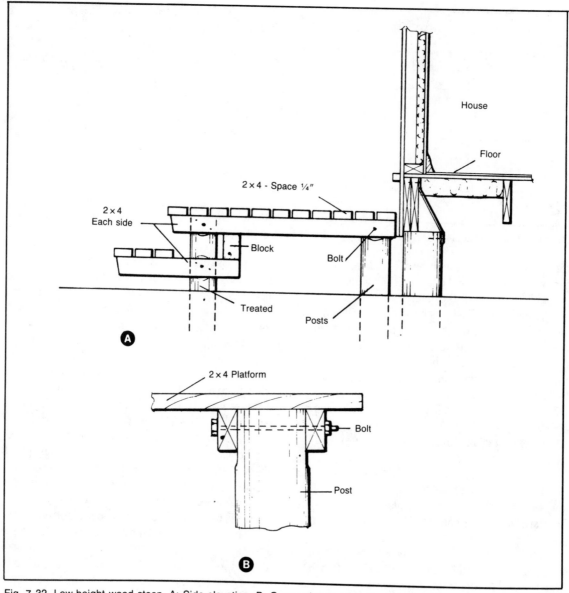

Fig. 7-32. Low-height wood stoop. A: Side elevation. B: Connection to post. Courtesy USDA

structed over a full basement or crawlspace normally require a platform with steps at outside doors. In houses with masonry foundations, these outside entry stoops consist of a concrete perimeter wall with poured concrete steps. However, a well-constructed wood porch will serve as well and usually costs less.

Inside stairs leading to an attic, second-floor bedrooms, or to the basement, must be provided for during construction of the house. This includes framing of the floor joists to accommodate the stairway and providing walls and carriages for the treads and risers. Even though the second floor might not be completed immediately, a stairway should be included during construction.

Outside Stoops

Outside stoops, platforms, and open plank stairs should give satisfactory service if these simple rules are followed:

■ Pressure-treat all wood in contact with or embedded in soil.

■ All untreated wood parts should have a 2-inch minimum clearance above the ground.

■ Avoid pockets or areas in the construction where water cannot drain away.

■ If possible, use wood having good decay resistance.

■ Use initial and regular applications of water-repellent preservative to exposed untreated wood surfaces.

■ Use vertical grain members.

A simple all-wood stoop consists of treated posts embedded in the ground, cross or bearing members, and spaced treads. One such design is shown in Fig. 7-32B. Because this type of stoop is low, it can serve as an entry for exterior doors where the floor level of the house is no more than 24 inches above the ground. Railings are not usually required. The platform should be large enough so that the storm door can swing outward freely. An average size is about 3½ feet deep by 5 feet wide.

Use treated posts 5 to 7 inches in diameter and embed them in the soil at least 3 feet. Nail and bolt (with galvanized fasteners) a cross member—usually a nominal 2-by-4-inch member) to each side of the posts (Fig. 7-32B). Block the inner ends to the upper cross pieces with a short piece of 2-by-4-inch member. Treads consist of 2-by-4-inch or 2-by-6-inch members, spaced about ¼ inch apart. Use two sixpenny galvanized plain or ring-shank nails for each piece at each supporting member.

Some species of wood have natural decay resistance and others need the application of a water-repellent preservative, followed by a good deck paint. If desired, a railing can be added by bolting short upright members to the 2-by-4-inch cross members. Horizontal railings can be fastened to the uprights.

A wood entry platform requiring more than one or two steps is usually designed with a railing and stair stringers. If the platform is about 3½ by 5 feet, two 1-by-12-inch carriages can be used to support the treads (Fig. 7-33A). In most cases, the bottoms of the carriages are supported by treated posts embedded in the ground or by an embedded treated timber. The upper ends of the carriages are supported by a 2-by-4-inch ledger fastened to posts and are face-nailed to the platform framing with twelvepenny galvanized nails. The carriage at the house side can be supported in the same way when interior posts are used.

When the platform is narrow, fasten a normal 3-by-5-inch ledger to the floor framing of the house with fortypenny galvanized spikes or 5-inch lag screws. Nail the 2-by-6-inch floor planks to the ledger and to the double 2-by-4-inch beam bolted to the post (Fig. 7-33B). Use two sixpenny galvanized plain or ring-shank nails for each tread. When a wide platform is desired, use an inside set of posts and double 2-by-4-inch or larger beams.

Railings can be made of 2-by-4-inch uprights bolted or lagscrewed to the outside beams. These members are best fastened with galvanized bolts or lag screws. Horizontal railing in 1-by-4- or 1-by-6-inch size can then be fastened to the uprights. When an enclosed skirting is desired, 1-by-4-inch slats can be nailed to the outside of the beam and an added lower nailing member. Treat all exposed untreated wood with a heavy application of water-repellent preservative. When a paint finish is desired, use a good deck paint.

Inside Stairs

When stairs to a second-floor area are required, the first-floor ceiling joists are framed to accommodate the stairway. When basement stairs are used, the first-floor joists must also be framed for the stairway. Two types of simple stair runs are commonly used in a small house: the straight run (Fig. 7-34A) and the long L (Fig. 7-34B). An open length of 10 feet is normally sufficient for adequate headroom, with a width of 2½ to 3 feet. A clear width of 2 feet, 8 inches is considered minimum for a main stair.

Two of the most important considerations in the design of inside stairways are headroom and the re-

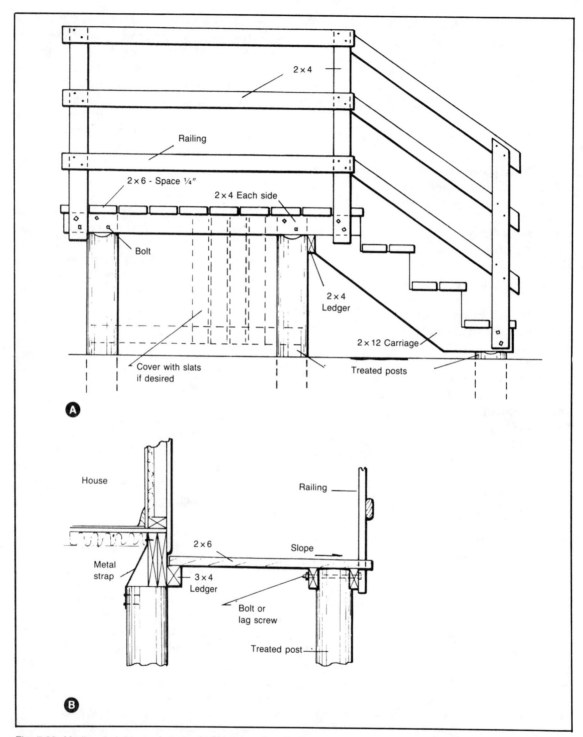

Fig. 7-33. Medium-height wood stoop. A: Side elevation; B: Connection to post. Courtesy USDA

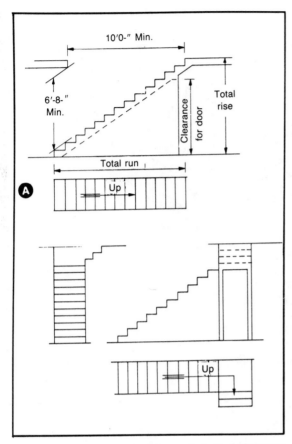

Fig. 7-34. Stair types. A: Straight run. B: Long L. Courtesy USDA

each stud (Fig. 7-38). The carriages can also be mounted directly to the wall studs or over the drywall finish, and the finish stringer then notched to them.

When no wall is present to fasten the tops of the carriages in place, use a ledger similar to that shown in Fig. 7-33A for an outside stair. The tops of the carriages are notched to fit this ledger. Two carriages are sufficient when treads are at least 1¹⁄₁₆-inch thick and the stair is less than 2 feet, 6 inches wide. Use three carriages when the stair is wider than this. When plank treads 1⅝-inch thick are used, two carriages are normally sufficient for stair widths up to 3 feet.

After carriages are mounted to the wall and treads and risers cut to length, nail the bottom riser

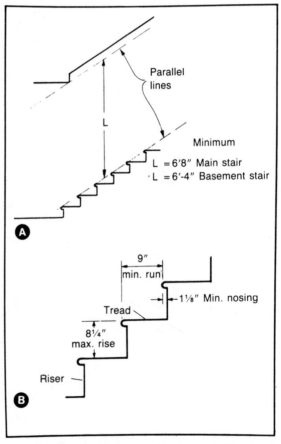

Fig. 7-35. Stair measurements. A: Head room. B: Riser-tread sizes. Courtesy USDA

lation of the riser height to the length of the tread. The minimum headroom for stairs should be 6 feet, 4 inches for main stairs (Fig. 7-35A). The relation of the riser to the run is shown in Fig. 7-35B. A good rule of thumb to apply is: The riser times the tread in inches should equal about 75.

When the length of the stairway is parallel to the joists, the opening is framed (Fig. 7-36). When the stairway is arranged so that the opening is perpendicular to the length of the joists, the framing should follow the details shown in Fig. 7-37.

The stair carriages are normally made from 2-by-12-inch members. They provide support for the stairs and nailing surface for the treads and risers. The carriages can be nailed to a finish stringer and to the wall studs behind with a sixteenpenny nail at

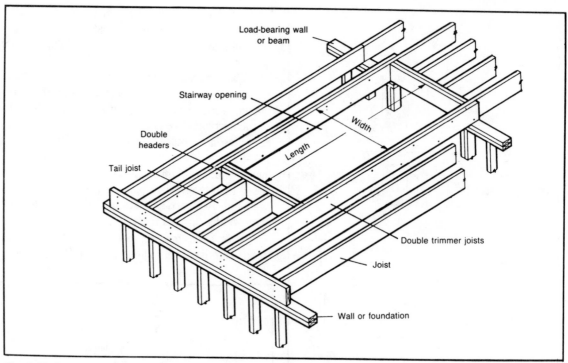

Fig. 7-36. Stairway parallel to joists. Courtesy USDA

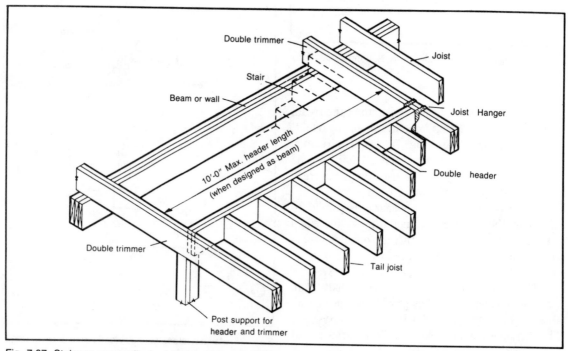

Fig. 7-37. Stairway perpendicular to joists. Courtesy USDA

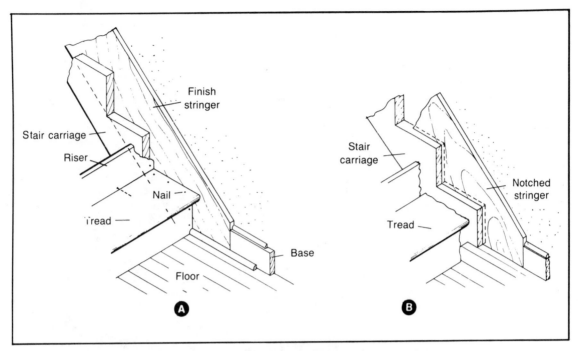

Fig. 7-38. Stair construction details. A: Full stringer. B: Notched stringer. Courtesy USDA

to each carriage with two eightpenny finish nails. The first tread, if $1\frac{1}{16}$-inch thick, is then nailed to each carriage with two tenpenny finish nails and to the riser below with at least two tenpenny finish nails. Proceed up the stair in this same manner. If $1\frac{5}{8}$-inch-thick treads are used, a twelvepenny finish nail might be required. Use three nails at each carriage, but eliminate riser below. All finish nails should be set.

In moderate-cost houses, finished stairs with stringers routed to fit the ends of the treads and risers, and with railing and balusters or a handrail, are generally used in main stairs to second-floor rooms. The safety factor of a handrail is essential.

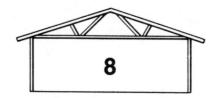

8

Paneling

Besides softwood plywood, the homeowner has several choices of panel products for the new home and, if you are adding on to your wood-frame home, paneling is one of the easiest products for the do-it-yourselfer to use. Some paneling provides needed strength and stiffness while other is used primarily for finish, sound reduction, or insulation. Take time to look them over and select the materials that best fit your need and pocketbook. There are many styles and colors from which to choose.

TYPES OF PANELING

Most of these products are partly or entirely of wood-based materials: hardwood plywood, insulation board, hardboard, laminated paperboard, particleboard, and gypsum board. Manufacturing and finishing methods vary greatly, which provides paneling materials with specific desirable properties. They are all manufactured in panel form, so they can cover large areas quickly and easily. Panel sizes are commonly 4 by 8 feet and can be handled by one man. Panels do, however, come in a range of sizes.

The most easily recognizable paneling is hardwood plywood. Veneers have been "unrolled" from the tree and the layers glued together to form a product that is particularly useful as decorative paneling for walls. Manufacturers often prefinish the panels.

Most other panel products are made from a process that involves breaking the wood down to small portions and then reassembling the elements into boards. For materials such as insulation board, hardboard, and laminated paperboard, the material is broken essentially into fibers, which are interfelted into panels. With particleboard, the wood is broken into particles that are bonded together with resin, heat, and pressure.

The other main member of the panel groups is gypsum board, which has a noncombustible gypsum core between faces of paper.

The insulation board-hardboard-paperboard group is customarily known as building fiberboard and may be called by such proprietary names as Celotex, Insulate, Masonite, Beaverboard, Homeasote.

Oldest of the boards is insulation board, made in two categories: semi-rigid and rigid. Semi-rigid consists of the low-density products used as insulation and cushioning. The semi-rigid type is used as insulation and cushioning. The rigid type includes both the interior board used for walls and ceilings, and the exterior board used for wall sheathing.

Hardboard is a grainless, smooth, hard product. It is used for siding underlayment and as prefinished fiberboards.

Particleboards are often known by the kinds of particle used in their makeup, such as flakeboard, chipboard, chipcore, or shavings board. New products have greater strength, stiffness, and durability than those made years ago. Some higher-quality products have been approved for subfloors and sheathing. Underlayment requires the largest amount of particleboard in houses, though much is also used for shelving.

Gypsum board is used principally for interior covering. Builders like it because, unlike plaster, it is a drywall material. Edges along the length are usually tapered to allow for a filled and taped joint. It may be obtained with a foiled back, which serves as a vapor barrier on the exterior walls, and is also available with a vinyl or other prefinished interior surface.

Each product will serve well if used as intended. Panels must be properly fastened to framing members and used in the right places under conditions for which they were designed. Manufacturers have found most instances of unsatisfactory performance directly related to improper application. Therefore, they usually include application and use instructions on each package or bundle of their material.

SPECIFIC JOBS

Let's talk about the paneling job you have in mind. Construction starts with the foundation and subflooring, proceeds to the wall and roof structure, and ends with the interior finish of the room.

Wall sheathing covers the outside wall framework of studs, plates, and headers, and ties them into a structural unit. It forms a flat base upon which the exterior finish can be applied. Many kinds of panel sheathing provide good strength and stiffness when properly fastened and thereby eliminate the need for corner bracing. Sheathing also serves to minimize air infiltration and, in certain forms, will provide good insulation.

Insulation board has been and continues to be used extensively for this purpose, reportedly more than any other residential construction sheathing material in the United States. It is inexpensive and it combines bracing strength, insulation, and a degree of noise control. Insulation board sheathing is available in three types: nail-base, intermediate density, and regular density—these in order of decreasing density and strength. Thickness is ½ inch in the first two types and ½ or 25/32 inch in the latter. Wood and asbestos shingles can be applied directly to the nail-base with annular-grooved nails. Regular density, as the least dense, is the best insulator.

The Federal Housing Administration recommends that all types be applied vertically on the wall. The boards provide adequate racking resistance without corner bracing, except that some specifying agencies require corner bracing with ½-inch regular-density board.

Insulation board and gypsum board sheathing are also available in 2-by-8-foot panels. They are applied horizontally and require corner bracing of the walls.

A special type of exterior particleboard sheathing panel, 4 feet wide, has been approved by some building codes. It must be no less than ⅜-inch thick and the studs must not be more than 16 inches on center. A ⅜-inch-thick sheathing grade of laminated paperboard is also approved. Corner bracing is not required for either.

Roof sheathing provides the strength and stiffness needed for expected loads, racking resistance to keep components square, and a base for attaching roofing. Panels have several advantages over lumber, particularly in ease and speed of application and resistance to racking.

Softwood plywood is now used most extensively for roof sheathing, but other panel materials are satisfactory if they meet performance requirements.

With wall sheathing in place, rough wiring and plumbing done, and insulation and vapor barriers installed as needed, you are ready to apply the inside ceiling and walls. Inside covering materials are generally considered as nonstructural. However, they usually contribute to the strength and stiffness of the ceiling or wall. Additionally, all must be capable of performing satisfactorily under the normal conditions to which they are subjected. Ceilings must not sag; walls must remain flat in use and withstand mild knocks and bumps.

Panel products are available in a wide variety of forms and finishes for interior coverings, and the homeowner generally finds these far easier to apply than lath and plaster. The thin sheet materials, however, require that studs and ceiling joists have good alignment to provide a smooth appearance.

Finishes in bathrooms and kitchens have more rigid stain and moisture requirements than those for other living spaces. Be especially careful when se-

lecting materials for these areas.

For ceilings, gypsum board is perhaps the most common sheet material, but large sheets of insulation board can also be used. The panels are generally nailed or screwed in place and then joints and fasteners are suitably covered or trimmed. Insulation board in the form of acoustical and decorative ceiling tile is popular, especially for dens and recreation rooms.

Interior wall coverings of many types are available. Panel materials most frequently applied are gypsum board, and hardboard that can be embossed or wood-grain printed.

Panels are usually 4 by 8 feet and applied vertically as single sheets so that the vertical joints butt at the stud. But there are many alternatives. For example, builders in some areas apply gypsum board

Fig. 8-1. Teton hardboard siding. Courtesy Georgia-Pacific Corp.

Fig. 8-2. Barnplank: weathered brown. Courtesy Georgia-Pacific Corp.

horizontally in sheet sizes 4 feet wide and up to 16 feet long, or they might use more than one layer.

Paneling can be used anywhere. It is probably the most versatile of all building products manufactured today and fits in beautifully with the popularity of do-it-yourself projects. There is really no mystery to installing wall paneling; it is an easy project. In new construction, if your walls are 8 feet high, simply follow the standard procedures. If you have a problem with walls in remodeling construction, you will find that there are many ways to overcome these problems with the use of paneling.

The color and type you choose will depend on your individual decorating desires. Certain panelings may be stained or painted, or in some areas, left to weather naturally (Figs. 8-1 and 8-2).

Western Red Cedar

One of the most popular paneling is western cedar, which for many years has played an important building role in the United States. Today contemporary styling is achieved by using tongue-and-groove, channel, v-groove, flush joint, and other patterns. Western red cedar is available in patterns with one side smooth and the other side rough- or saw-textured. Western red cedar has natural resistance to decay, which gives it uncommon durability. Here are some more of its desirable characteristics:

■ Dimenional Stability — approved testing shows that western red cedar has the lowest volumetric shrinkage of any of the domestic species.

209

■ Insulation—extremely low density and fine closed cellular structure makes it an excellent heat and cold transfer barrier.

■ Weather Resistance—applied properly, western red cedar affects minimal dimensional disturbance and resists pulling loose from fastenings when exposed to the elements.

■ Working Qualities—It is one of the easiest species of softwoods to work with hand tools. Its gluing capabilities are rated extremely high both with various adhesives and gluing conditions and it is highly resistant to splitting.

■ Weight: western red cedar is one of the lightest of any commercial species, making it easy to handle.

Handle western red cedar bevel siding properly: keep it dry and store under cover as soon as it is received. And, as with any product of value, avoid rough handling in storage and on the job.

INSTALLATION

The basics of proper siding installation must include use of good nails and accepted nailing practices. Recommended nail types are as follows:

■ Galvanized—hot-dipped zinc coating provides outstanding corrosion resistance.

■ Aluminum—corrosive-resistant; will not discolor or deteriorate the wood siding.

■ Stainless steel—very satisfactory for the job.

There are numerous designs of nail heads, from flat (or "sinker") to casing and finishing types. These are all effective. Increased holding power may be obtained by using ring-threaded nail shanks. Smooth-shanked nails tend to loosen under extremes in temperature. As for nail points—blunt points are very effective in holding; a sharp point tends to cause splitting. For best holding with minimum splitting, use blunt, medium diamond, or medium needle—all with a ring-threaded shank.

The following are application and nailing procedures for various types of siding.

Bevel Siding: Size of nail depends on the material being used. For bevel siding to 8 inches wide

(½ or ⅝-inch butt), use sixpenny nails. For 8-inch and wider bevel siding (¾-inch butt), an eightpenny nail is recommended. A 1-inch lap is customary in applying bevel siding.

For regular bevel siding, drive the nail slightly more than 1 inch above the thick edge of the pieces. It is important to just clear the thin edge of the underlying piece on each stud crossing. Tap the nail flush with the siding surface or countersink then putty over it. Nails should be long enough to penetrate into studs (or stud and sheathing combined) at least 1½ inch.

Tongue-and-Groove Siding: Nail siding directly to framing in both horizontal and vertical applications. Blocking 24 inches on center is required for vertical application of ½-inch tongue-and-groove siding. For ¾-inch tongue-and-groove siding, 48-inch on center blocking will suffice. Narrow widths to 6 inches are blind-nailed, one nail per bearing through tongue. Use hot-dipped galvanized, mechanically plated, or aluminum nails of high tensile strength. For patterns 8 inches and wider, face-nail with one eightpenny rust-resistant siding nail per bearing on exterior applications.

In exterior applications, when two lengths of vertical siding are joined with a 45-inch bevel cut at the end joint, moisture penetration is eliminated. The tip of the top piece extends over the lip of the lower piece, which forces moisture down the 45-degree angle toward the exterior surface.

Board and Batten Siding: Fasten siding with one rust- and corrosion-resistant eightpenny siding nail per bearing, driven through the center of the board. Space about ½ inch. Fasten batten with one tenpenny siding nail per bearing driven through the center of the batten strip, so that the shank passes between the underlying boards (Fig. 8-3).

Channel Groove Siding: Face nailing is recommended, but do not nail through the underlying joint because expansion and contraction would then be retarded. Here are some suggestions when using unseasoned materials: Make allowances for shrinkage; use as narrow a width as possible and use products that allow for some degree of shrinkage (board and batten, channel rustic with an adequate tongue, board on board, narrow rustic bevel, etc.).

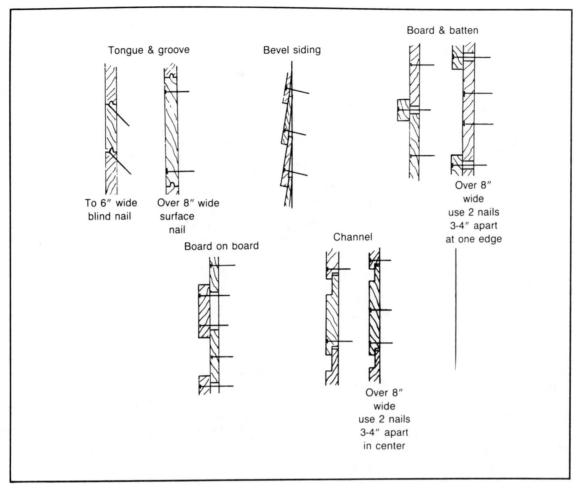

Tongue & groove

To 6" wide
blind nail

Over 8" wide
surface
nail

Bevel siding

Board & batten

Over 8"
wide
use 2 nails
3-4" apart
at one edge

Board on board

Channel

Over 8"
wide
use 2 nails
3-4" apart
in center

Fig. 8-3. Paneling grades. Courtesy Western Wood Products

Nail properly and season in place before applying finish or pre-stain.

Fiberboard Underlayment: In every case, when applying siding material over fiberboard sheathing, it is important that nails be driven into the studding for solid support.

Figure 8-4 provides Exterior Wall Detail.

Applying Interior Panels

The first panel is the important one. Put the first panel in place and butt to adjacent wall in the corner. Make sure it is completely plumb and that both left and right panel edges fall on solid stud backing (Fig. 8-5). Most corners are not perfectly true, how-ever, so you will probably need to trim the panel to fit into the corner properly.

Fit the panel into the corner, checking with a level to be sure the panel is plumb vertically. Draw a mark along the panel edge, parallel to the corner. On rough walls like masonry, or adjoining a fireplace wall, scribe or mark the panel with a compass on the inner panel edge, then cut on the scribe line to fit (Fig. 8-6).

Scribing and cutting the inner panel edge might also be necessary if the outer edge of the panel does not fit directly on a stud. The outer edge must fall on the center of a stud to allow room for nailing your next panel. Before installing the paneling, paint a stripe of color on the wall where panels will meet

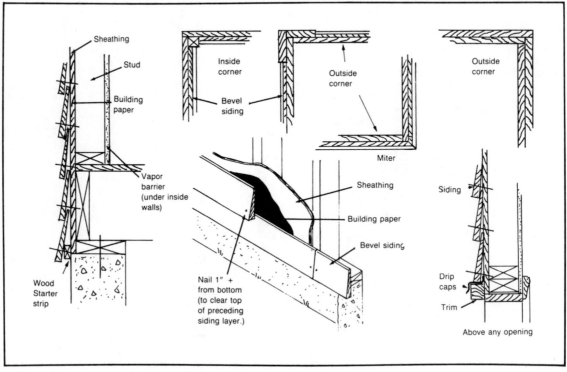

Fig. 8-4. Exterior wall detail. Courtesy Western Wood Products

that matches the color on the paneling groove. The appearance of any slight gap between the panel edges will be minimized in this way.

Paneling is usually grooved every 16 inches. This allows most nails to be placed in the grooves, falling directly on the 16-inch stud spacing. Regular small-headed finish nails or colored paneling nails can be used.

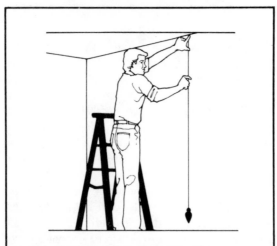

Fig. 8-5. Plumb first panel. Courtesy American Plywood Assn.

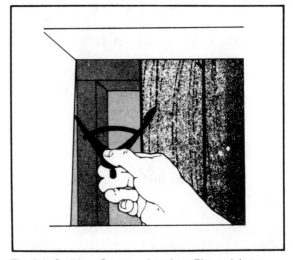

Fig. 8-6. Scribing. Courtesy American Plywood Assn.

Fig. 8-7. Nail spacing. Courtesy American Plywood Assn.

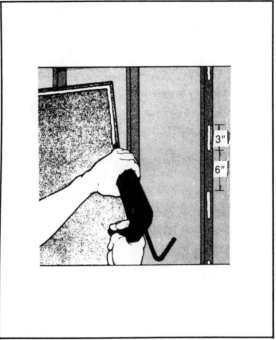

Fig. 8-8. Caulking with adhesive. Courtesy American Plywood Assn.

For paneling to studs, threepenny (1½-inch) nails are recommended, but if you must penetrate backer board, plaster, or drywall, use sixpenny (2-inch) nails to give a solid bite into the stud. Space nails 6 inches apart on the panel edges and 12 inches apart in the panel field (Fig. 8-7). Nails should be countersunk slightly below the panel surface with a nail set, then hidden with a matching putty stick. Colored nails eliminate the need to countersink and putty. Use 1-inch colored nails to apply paneling to studs, 1⅝-inch nails to apply paneling through gypsum board or plaster.

Using adhesive to install paneling is a simple method that eliminates the chore of countersinking and hiding nail heads. Adhesive may be used to apply paneling directly to studs or over existing walls, as long as the surface is sound and clean.

Paneling must be properly cut and fitted prior to installation. Make sure the panels and walls are clean, and free from dirt and particles before you start. Once applied, the adhesive makes adjustments difficult. However, used properly, adhesive gives a more professional appearance.

A caulking gun with adhesive tube is the simplest method of application. Trim the tube end so that a ⅛-inch-wide adhesive bead can be squeezed out. Once the paneling is fitted, apply beads of adhesive in a continuous strip along the top, bottom, and both edges of the panel. On intermediate studs, apply beads of adhesive 3 inches long and 6 inches apart (Fig. 8-8).

With scrap plywood or shingles used as a spacer at the floor level, set the panel carefully in place and press firmly along the stud lines, spreading the adhesive into the wall. Then using a hammer with a padded wooden block or a rubber mallet, tap over the glue lines to assure a sound bond between the panel and backing.

Be sure to read the adhesive manufacturer's instructions carefully prior to installation. Some require the panel to be placed against the adhesive, then gently pulled away from the wall for a few minutes. This allows the solvent to "flash off" and the adhesive to set up. The panel is then repositioned and tapped home.

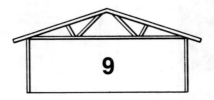

9

Fireplaces

Fireplaces have become more efficient in the 80s than they were years ago. With the new technology there are built-in (prefabricated) units and freestanding units that give not only as visual comfort from the beauty of a burning fire on a cool winter day, but also aid in savings on your fuel bill.

A fireplace consists of a variety of parts, not the least important of which, of course, is the chimney. Chimneys are generally constructed of masonry units supported on a suitable foundation. A chimney must be structurally safe, and capable of carrying away harmful gases from the fuel-burning equipment and other utilities. Lightweight, prefabricated chimneys that do not require masonry protection or concrete foundations are now accepted for certain uses. Make certain, however, they are UL-(Underwriters Laboratories) approved.

In addition to being safe and durable, fireplaces should also be constructed so that they provide sufficient draft and are suitable for their intended use.

Improved heating efficiency and the assurance of a correctly proportioned fireplace can be obtained by installing a factory-made circulating insert into a fireplace opening. This metal unit, enclosed by the masonry, allows air to be heated and circulated throughout the room in a system separate from the direct heat of the fire.

Certain wood stoves will allow an attachment for heating water. These can save on fuel and are good in emergency situations and during power failures.

CHIMNEYS

The chimney should be built on a concrete footing of sufficient area, depth, and strength for the imposed load. The footing should be below the frostline. For houses with a basement, the footings for the walls and fireplace are usually poured together and at the same elevation.

The size of the chimney depends on the number of flues, and the design of the house. The house design could include a room-wide brick or stone fireplace wall that extends through the roof. While only two or three flues might be required for heating units and fireplaces, several "false" flues may be added at the top for appearance.

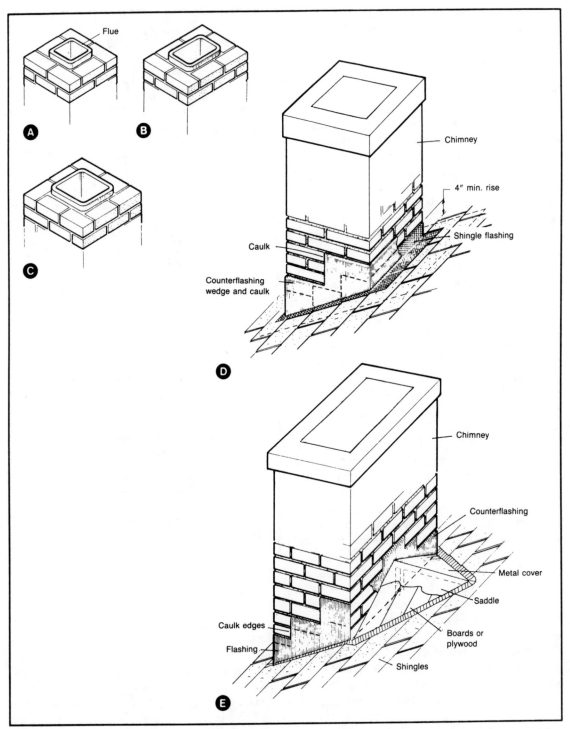

Fig. 9-1. Brick and flue combinations and chimney flashing: A: 8 by 8 flue lining. B: 8 by 12 flue lining. C: 12 by 12 flue lining. D: Flashing without saddle. E: Chimney saddle. Courtesy USDA

The flue sizes are made to conform to the width and length of a brick so that full-length bricks can be used to enclose the flue lining. Thus an 8½-by-8-inch flue lining (about 8½ by 8½ inches in outside dimensions) with the minimum 4-inch thickness of surrounding masonry will use six standard bricks for each course (Fig. 9-1). An 8-by-12 inch flue lining (8½ by 13 inches in inside dimensions) will be enclosed by seven bricks at each course, a 12-by-12 inch flue (13 by 13 inches in outside dimension) by eight bricks, and so on. Each fireplace should have a separate flue and, for best performance, flues should have a 4-inch-wide brick spacer (withe) between them (Fig. 9-2).

The greater the difference in temperature between chimney gases and the outside atmosphere, the better the draft. An interior chimney will have better draft because the masonry will retain heat longer. The height of the chimney as well as the size of the flue are important in providing sufficient draft.

The height of a chimney above the roofline usually depends upon its location in relation to the ridge. The top of the extending flue liners should not be less than 2 feet above the ridge or a wall that is within 10 feet. For flat or low-pitched roofs, the chimney should extend at least 3 feet above the highest point of the roof. To prevent moisture from entering between the brick and flue lining, a concrete cap is usually poured over the top course of brick. Precast or stone caps with a cement wash are also used.

Flashing for chimneys was illustrated in Fig. 9-1. Masonry chimneys should be separated from wood framing, subfloor, and other combustible materials. Framing members should have at least a 2-inch clearance and be firestopped at each floor with asbestos or other types of noncombustible material (Fig. 9-3). Subfloor, roof sheathing, and wall sheathing should have a ¾-inch clearance. A cleanout door is included in the bottom of the chimney where there are fireplaces or other solid fuel-burning equipment, and at the bottom of other flues. The cleanout door for the furnace flue is usually located just below the smokepipe thimble with enough room for a soot pocket.

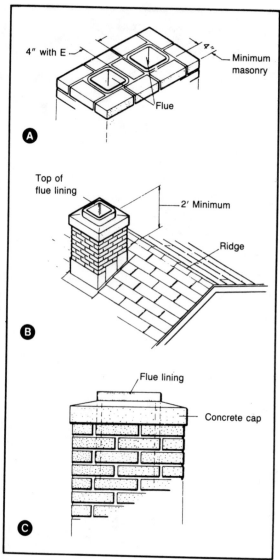

Fig. 9-2. Chimney details. A: Spacer between flues. B: Height of chimneys. C: Chimney cap.

Flue Linings

Rectangular flue linings of fire-clay or round vitrified tile are used in chimneys. Vitrified (glazed) tile or a stainless lining is usually required for gas-burning equipment. Local codes outline these specific requirements. A fireplace chimney with at least an 8-inch-thick masonry wall ordinarily does not

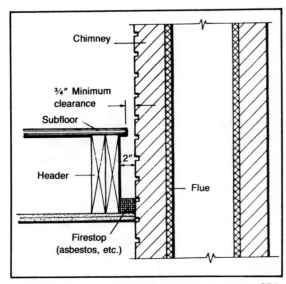

Fig. 9-3. Clearances for wood construction. Courtesy USDA

require a flue lining. However, the cost of the extra brick or masonry and the labor involved (if you do not do it yourself) are most likely greater than the cost of the flue lining. Furthermore, a well-installed flue lining will result in a safer chimney.

Install flue liners ahead of the brick or masonry work. As it is carried up, careful bedding of the mortar will result in a tight and smooth joint. When diagonal offsets are necessary, bevel the flue liners at the direction change in order to have a tight joint. It is also good practice to stagger the joints in adjacent tile.

Flue lining is supported by masonry and begins at least 8 inches below the thimble for a connecting smoke or vent pipe from the furnace. In fireplaces, the flue liner should start at the top of the throat and extend to the top of the chimney.

Rectangular flue lining is made in 2-foot lengths and in sizes of 8 by 8, 8 by 12, 12 by 12, 12 by 16, and up to 20 by 20 inches. Wall thicknesses of the flue lining vary with the size of the flue. The smaller sizes have a ⅝-inch-thick wall, and the larger sizes vary from ¾ inch to 1⅜ inches in thickness. Vitrified tiles 8 inches in diameter are most commonly used for the flues of the heating unit, although larger sizes are also available. This tile has a bell joint.

Maintenance

Chimneys should be inspected every fall for defects. Check for loose or fallen bricks, cracked or broken flue lining, and excessive soot accumulation, by lowering an electric light into the flue. Mortar joints can be tested from the outside by prodding with a knife.

If inspection shows defects that cannot be readily repaired or reached for repair, you should tear the masonry down and rebuild properly. Do not use old bricks that have been impregnated with soot and creosote in the new work, because they will stain plaster whenever dampness occurs. Soot and creosote stains are almost impossible to remove.

Chimney cleaning should be performed periodically. Hiring a commercial cleaning firm is one way to accomplish this. Chemical soot removers can also be used.

If there is not too great an offset in the chimney, you can dislodge soot and loose materials by pulling a weighted sack of straw up and down in the flue. Seal the front of a fireplace when cleaning the flue to keep soot out of the room.

Chimneys must also be cleaned of creosote, a black, carbon-based substance that builds up inside the chimney. This substance can ignite and cause chimneys and even entire homes to be burned and ruined. Pulling wire brushes up and down inside the chimney helps to dislodge creosote. Certain woods, such as pine, cause creosote more than others.

THE FIREPLACE

A well-designed, properly built fireplace has many benefits. It can provide additional heat in cold climates, and provide all the heat necessary in mild climates. It enhances the appearance and comfort of any room in any house in any climate. A fireplace can also burn as fuel certain combustible materials that otherwise might be wasted—for example, coke, briquets, and scrap lumber.

A fireplace adds to the attractiveness of the house interior, but one that does not "draw" properly is a detriment, not an asset. By following several rules concerning the relation of the fireplace opening size to flue area, depth of the opening, and other measurements, satisfactory performance can be assured. Metal circulating fireplaces, which form the main outline of the opening and are enclosed with brick, are designed for proper functioning when flues are the correct size.

One rule is that the depth of the fireplace should be about two-thirds the height of the opening. Therefore, a 30-inch-high fireplace would be 20 inches deep from the face to the rear of the opening.

The flue area should be at least one-tenth of the open area of the fireplace (width times height) when the chimney is 15 feet or more in height. When less than 15 feet, the flue area in square inches should be one-eighth of the opening of the fireplace. This height is measured from the throat to the top of the chimney. A fireplace with a 30-inch width and a 24-inch height (720 square inches) would require an 8-by-12-inch flue, which has an inside area of about 80 square inches. A 12-by-12-inch flue liner has an area of about 125 square inches, and this would be large enough for a 36-by-36-inch opening when the

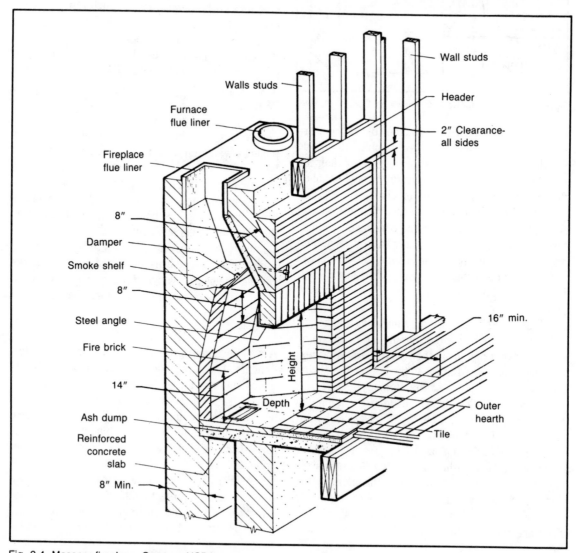

Fig. 9-4. Masonry fireplace. Courtesy USDA

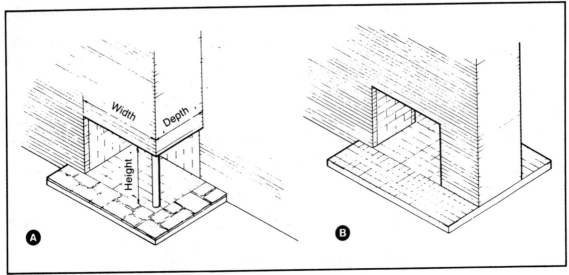

Fig. 9-5. Dual-opening fireplace; A, Adjacent opening; B, through fireplace. Courtesy USDA

chimney height is 15 feet or over.

The back width of the fireplace is usually 6 to 8 inches narrower than the front. This helps to guide the smoke and fumes toward the rear. A vertical backwall about 14 inches high then tapers toward the upper section or "throat" of the fireplace (Fig. 9-4). The area of the throat should be about 1¼ to 1⅓ times the area of the flue to promote better draft. Use an adjustable damper at this area for easy control of the opening.

The smoke shelf (top of the throat) is necessary to prevent back drafts. The height of the smoke shelf should be 8 inches above the top of the fireplace opening. The smoke shelf is concave to retain any slight amount of rain that might enter.

Steel angle iron is used to support the brick or masonry over the fireplace opening. The bottom of the inner hearth, the sides, and the back are built of a heat-resistant material such as firebrick. The outer hearth should extend at least 16 inches out from the face of the fireplace and be supported by a reinforced concrete slab. This outer hearth is a precaution against flying sparks and is made of noncombustible materials such as glazed tile. Hangers and brackets for fireplace screens are often built into the face of the fireplace.

Fireplaces with two or more openings require much larger flues than the conventional fireplace. For example, a fireplace with two open adjacent faces (Fig. 9-5A) would require a 12-by-16-inch flue for a 34-by-20-by-30-inch (width, depth, and height, respectively) opening. Local building regulations usually cover the proper sizes for these types of fireplaces.

Location of Fireplace

Location of the fireplace within a room depends on the location of the existing or proposed chimney. A fireplace should not be located near doors.

The most common size for fireplace openings are from 2 to 6 feet wide. The kind of fuel to be burned can suggest a practical width. For example, where cordwood (4 feet long) is cut in half, an opening 30 inches wide is desirable; where coal is burned a narrower opening can be used. Height of the opening can range from 24 inches for an opening 2 feet wide to 40 inches for one that is 6 feet wide. The higher the opening, the greater the chance of a smokey fireplace.

In general, the wider the opening, the greater the depth. A shallow opening throws out relatively more heat than a deep one, but holds smaller pieces of wood. You have the choice, therefore, between a deeper opening that holds larger, longer-burning logs and a shallower one that takes smaller pieces of wood, but throws out more heat. In small fireplaces, a depth of 12 inches might permit good draft, but a minimum depth of 16 inches is recommended to lessen the danger of brands falling out on the floor.

Suitable screens should be placed in front of all fireplaces to minimize the danger from brands and sparks. Glass enclosures are especially desirable since they are ideal for safety and attractive. They can be closed at night so the heat loss that would go up the chimney from the "cold" fireplace is able to be cut off.

Second-floor fireplaces are usually made smaller than first-floor ones, because of the reduced flue height.

Construction

Fireplace construction is basically the same regardless of design. Figure 9-6 shows a typical fireplace.

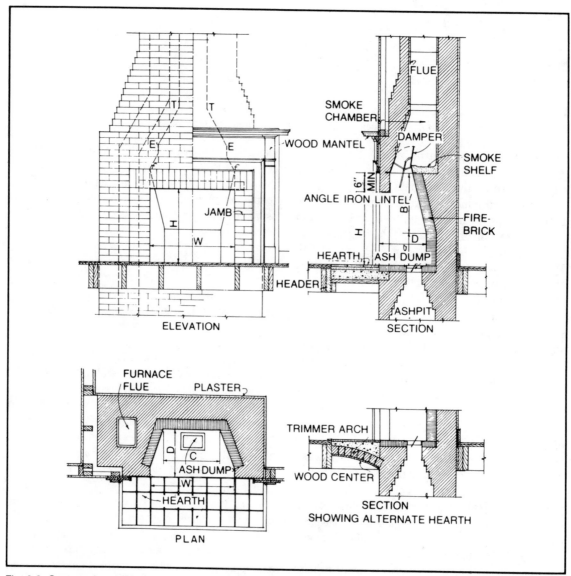

Fig. 9-6. Construction details of a typical fireplace. Courtesy USDA

Table 1-9 provides recommended dimensions for essential parts of fireplaces of various sizes.

Footings: Foundation and footing construction for chimneys with fireplaces is similar to that for chimneys without fireplaces. Be sure the footings rest on good firm soil below the frostline.

Hearth: The fireplace hearth should be made of brick stone, terra cotta, or reinforced concrete at least 4 inches thick. It should project at least 20 inches from the chimney breast and should be 24 inches wider than the fireplace opening (12 inches on each side). The lower right-hand drawing in Fig. 9-6 shows an alternate method of supporting the hearth.

In buildings with wooden floors, the hearth in front of the fireplace should be supported by masonry trimmer arches or other fire-resistant construction. Remove wood centering under the arches used during construction of the hearth and hearth extension when construction is completed.

The hearth can be flush with the floor so that sweepings can be brushed into the fireplace, or the hearth can be raised. Raising the hearth to various levels and extending in length as desired is presently common practice, especially in contemporary design (Fig. 9-7). If there is a basement, a convenient ash dump can be built under the back of the hearth. An ashpit for a fireplace should be of tight masonry and

Fig. 9-7. Well-designed contemporary fireplace. Courtesy USDA

should be provided with a tightly fitting iron cleanout door, and a frame 10 inches high and 12 inches wide (Fig. 9-8).

The recommended method of installing floor framing around the hearth is shown in Fig. 9-9. Where a header is more than 4 feet long, it should be doubled as shown. If it supports more than four

Table 9-1. Recommended Dimensions for Fireplaces.

Size of fireplace opening			Minimum width of back wall	Height of vertical back wall	Height of inclined back wall	Size of flue lining required	
Width	Height	Depth				Standard rectangular (outside dimension)	Standard round (inside diameter)
w	h	d	c	a	b		
Inches	Inches	Inches	Inches	Inches	Inches	Inches	Inches
24	24	16-18	14	14	16	8½ × 13	10
28	24	16-18	14	14	16	8½ × 13	10
30	28-30	16-18	16	14	18	8½ × 13	10
36	28-30	16-18	22	14	18	8½ × 13	12
42	28-32	16-18	28	14	18	13 × 13	12
48	32	18-20	32	14	24	13 × 13	15
54	36	18-20	36	14	28	13 × 18	15
60	36	18-20	44	14	28	13 × 18	15
54	40	20-22	36	17	29	13 × 18	15
60	40	20-22	42	17	30	18 × 18	18
66	40	20-22	44	17	30	18 × 18	18
72	40	22-28	51	17	30	18 × 18	18

Courtesy USDA

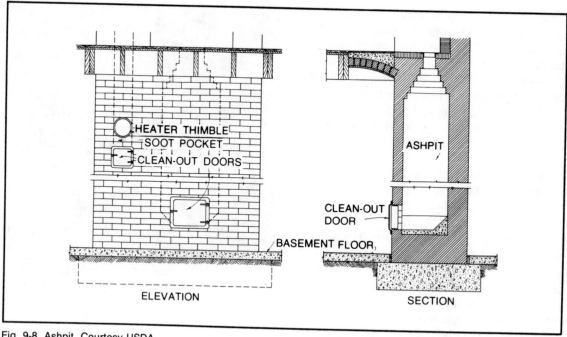

Fig. 9-8. Ashpit. Courtesy USDA

tall beams, its ends should be supported in metal joist hangers. The framing may be placed ½ from masonry chimney walls 8 inches thick.

There are other parts of the fireplace that should be taken into consideration: the jambs, lintel, throat, and damper—all of which are outlined in the above figures.

Do-It-Yourself Fireplaces

Many prefabricated models of fireplace inserts are available in today's modern market. For the do-it-yourselfer, summer is the ideal time to enhance the beauty of your home and get ready for winter by installing a modern, factory-built fireplace insert. Cost of these is far less than a masonry unit and installation is relatively simple.

A factory-built fireplace insert opens new decorating potentials because it can be installed almost anywhere. It is, of course, best to install your fireplace at the time of initial construction. If you cannot do this, due to cost or any other circumstance, do not give up the idea for the future. Take your time, learn about the various units available, shop around at your local hardware stores.

The factory-built insert in Fig. 9-10 can be completely installed in just a few hours. This unit is for masonry openings only. It is highly efficient as a forced-air blowing unit and will give you years of efficient, high-quality fireplace service. Plus, it is attractive.

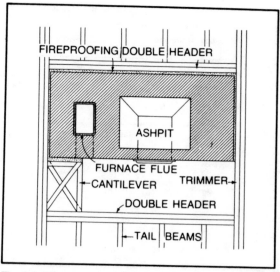

Fig. 9-9. Floor framing around hearth. Courtesy USDA

222

Fig. 9-10. Ponderosa fireplace insert. Courtesy Birmingham Stove & Range Company

Safe installation of any prefabricated built-in insert or free-standing model is imperative. Do not install units such as the one illustrated in Fig. 9-10 in a metal prefabricated, zero-clearance, manufactured, or factory-built fireplace. Some of these fireplaces might be faced with brick, but the inside of the fireplace will not be made of brick. These fireplaces are made for installation near materials that can catch fire and so they use air flowing through the firebox to cool the fireplace. Your insert could interrupt this cooling air and could cause an increased risk of fire in the walls and floor of your home. In such cases, install your insert only in a masonry (brick or stone) fireplace. If you have any doubt about the materials used in the construction of your current fireplace, get a qualified mason to inspect it before installing your factory-built.

Your insert blower is designed to use a grounded electrical wall socket. Do not change the insert's plug. If your wall socket is not made to ac-

cept a three-prong plug, you must use an adapter with the plug and attach the adapter's green ground wire to the screw on the wall plug cover. Make sure the cover screw is connected to a ground. An electrician can inspect the wall socket and tell you if it is grounded. If you do not ground the insert you could receive an electrical shock. This system is designed for providing heat most efficiently into rooms beyond that in which your fireplace is located.

Use your fireplace safely. Build fires only inside the insert, using the log grate that comes with the insert. Keep the fire behind the front prongs of the grate.

Maintenance of a prefabricated unit is not difficult, but it is important to assure continual efficiency and safety. Keep ashes removed from your insert. Large amounts of ashes can block the draft opening and cause an increased risk of fire. Keep the blower grill and the blower outlets clean so air can circulate around the fire in the insert.

You must keep your chimney clean, as outlined in the maintenance section on chimneys. Burning wood produces a sticky "goo" called creosote. This material is deposited on the inside of the chimney by wood smoke. It catches fire easily and can cause a chimney fire, which is very dangerous.

A protective floor covering will be required for all combustible floors.

Assembly of Insert

Before beginning to assemble or install your insert, be certain that the masonry fireplace used with the insert is of the proper minimum size. Make sure damper handles in the fireplace won't be in the way of the insert.

Do not use the insert in a fireplace that does not meet these minimum requirements. If you do, you may suffer property damage, personal injury, or even death. Be sure to discuss your fireplace opening with the manufacturer or manufacturer's representative prior to purchasing any prefabricated unit for insert into your fireplace opening. Figure 9-11 shows the minimum masonry fireplace dimensions.

The tools you will need to install your insert are: an adjustable wrench, a Phillips screwdriver, a flat-blade screwdriver, electrical tape, pliers, a small hammer, rule or tape, and scissors or a knife.

Right Door Assembly: Refer to Fig. 9-12. Put threaded portion of handle (6) through right door and hold at the 4 o'clock position. Slip ½-inch washer on threaded end (1). Place latch (8) hole over the square part of handle behind the threaded portion and make sure the latch is pointing toward the 6 o'clock position. Place a 5/16-inch flat washer (10) and a 5/16-inch (9) lock washer on the threaded portion and follow with a 5/16-inch hex nut (12). Holding handle in 4 o'clock position, securely fasten using wrench—but do not overtighten.

Left Door Assembly: Refer to Fig. 9-13. Put threaded portion of handle (6) through the left door and hold at the 7 o'clock position. Slip 3½-inch washers (11) on threaded end. Follow with a 5/16-inch flat washer (10) and a 5/16-inch hex nut (12). Holding handle in 7 o'clock position, tighten securely. Do not overtighten.

Changing the Power Cord: Refer to Fig. 9-14. Stand facing your fireplace. If the nearest electrical wall socket is on the left side, it will be necessary to change the cord to the left side of the insert. If the wall socket is on the right side of your masonry fireplace, you will not need to follow the directions for the section "Assembly of Enclosure," which follows.

Make sure the insert is unplugged from power before beginning. Remove the fan cover plate by

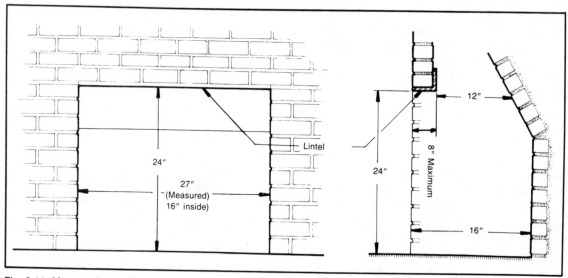

Fig. 9-11. Masonry fireplace minimum dimensions. Courtesy Birmingham Stove & Range Company

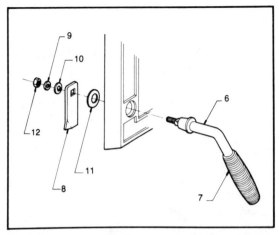

Fig. 9-12. Right door handle assembly. Courtesy Birmingham Stove & Range Company

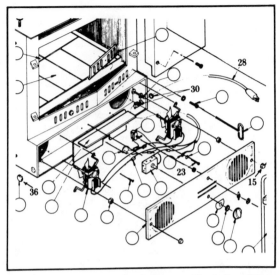

Fig. 9-14. Fireplace insert detail. Courtesy Birmingham Stove & Range Company

removing the two acorn nuts (15) from the right and left sides of the cover plate below doors. Remove the self-tapping screw (23) from the lower left motor leg to free the green ground lead and connector. Reinstall washer and screw.

Using a flat-blade screwdriver, open the quick splice connectors and disconnect the white wire of the power cord (28) and the white wire of the motor control. Also disconnect the black wire of the power cord and the black wire of the blower fan. Tape the point where connector pierced the wire with electrical tape.

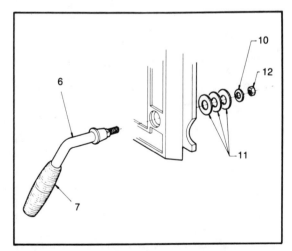

Fig. 9-13. Left door handle assembly. Courtesy Birmingham Stove & Range Company

Loosen two screws on strain relief bushing (30), clamping the power supply cord. Pull cord carefully through bushing. Loosen bushing nut and remove bushing from hole.

Using a screwdriver, pry the plug (36) from the hole on the left side of the insert cabinet. Reinstall by tapping into the hole where the strain relief bushing was located on the right side of the cabinet.

Install strain relief bushing with screws in hole in left side of cabinet. The cord clamp must be on the inside of the cabinet. Reinsert power cord and tighten clamp screws.

Follow these wiring instructions carefully or the motor control may be burned out (Fig. 9-15): Remove the tapping screw from the left motor's lower right mount leg. Slip the green ground wire and connector on the screw shaft and reinstall screw. Tighten securely. No wire stripping is necessary with the quick-splice connector. Using the connector, join the white wire of the power cord to the white wire leading from the motor control.

Use the other quick-splice to join the black wire of the power cord to the black wire with red sleeve leading from the left side of the left motor (No. 1) as viewed from the front of the insert. Notice this black motor wire is connected to the wire from the right motor which the power cord was previously

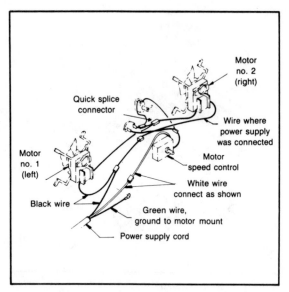

Fig. 9-15. Left side electrical wiring detail. Courtesy Birmingham Stove & Range Company

Installing the Insert

Before beginning your installation, consult your applicable code clearance requirements. The minimum clearances listed are most widely accepted nationally.

With the insulation in place, the power cord changed (if needed), and the door handles installed, you are ready to install the insert in your masonry fireplace and make sure the insert will fit. Be sure the chimney is cleaned and not blocked.

Prior to installation of the masonry fireplace, make sure the damper is opened. To avoid smoke damage, secure the damper handle or damper so that it cannot accidentally shut. Make sure the damper handle and any other object in the fireplace will be clear of the insert when it is installed.

The floor of the fireplace must be level. If it is not, the enclosure might not fit properly. Use a material that will not catch fire, if needed, to level the fireplace floor. You might also want to remove the

spliced to. Attaching the black power cord wire to the other black wire of the motor will cause the motor control to be burned out.

After wiring, replace the fan cover plate with acorn nuts.

Assembly of Enclosure

Refer to Fig. 9-16. Place the top and side plates on a smooth surface (preferably a rug so the enclosure finish will not be scratched) with the flanges up. Line up the two side plates and attach to the top plate using the four 10-by-½-inch tapping screws.

Place the fireplace insulation (43) on the outside edges of the enclosure plates. Place the double-stick tape on the enclosure, remove the backing, and press in the fiberglass insulation—foil side to the tape. It will be easier if you roll the backing back a little at the time and firmly press the insulation to the tape. Use scissors to cut the tape and the fiberglass.

Next, attach the enclosure clips (55) to the body of the insert. Place the ⅜-inch washers (53) onto the 6 studs (two each on each side of the top) of the insert body. Place the slot in the clips on the studs and then place the ¼-inch washer (38) and the ¼-inch hex nuts (37) on the studs. Tighten securely.

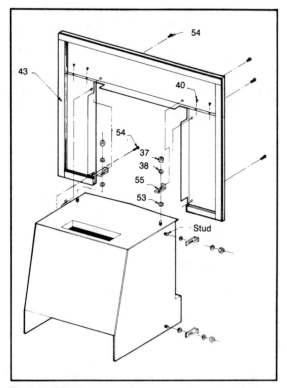

Fig. 9-16. Enclosure assembly, rear view. Courtesy Birmingham Stove & Range Company

doors and firebrick from the inside of the insert to reduce the weight of the insert.

You will need assistance in installing the insert because it is very heavy.

Slide the insert back into the fireplace. It will be easier if you place a ½-inch section of pipe under the insert; otherwise the fireplace brick could get damaged.

Double check to see that the smoke outlet of the insert is not blocked by the lintel of the fireplace. Do this by placing your hand through the insert smoke outlet or check with a light. Make sure the opening into the chimney is over the insert smoke outlet. Check to see that the insert is level. Wedge the bottom of the insert if necessary with any non-combustible material so that the top is level.

With the insert all the way into the masonry fireplace, position the enclosure so that: the holes in the front of the top and side plates line up with the holes in the enclosure clips; the flanges on the plates slide under the enclosure clips; and that the insulation on the back of the enclosure is pressed firmly against the outside of the fireplace. Adjust the clips if needed.

If doors and brick were removed, reinstall now.

Plug the blower power cord into a convenient wall socket. *Important*: The blower system is designed for use in a grounded wall socket. Do not change the insert's plug. If the wall socket is not made to accept a three-prong plug, you must use an adapter with the plug and attach the adapter's green ground wire to the screw on the wall socket cover. Make sure the cover screw is connected to ground. An electrician can inspect the socket and tell you if it is grounded. If you do not ground the insert's electrical system you could be shocked or even electrocuted.

COAL-BURNING STOVES

Whether it is best to buy a coal-burning stove in lieu of a wood burning stove depends, in part, upon the area of the country in which you are located. Which is best?

There are advantages to using coal instead of wood. Conversely, there are just as many advantages to using wood over coal. If you live in the southern or western United States, certainly wood is more readily available. Stove owners must select the fuel—and the stove—that's best for their needs. The cast-iron, coal-burning stove in Fig. 9-17 is an energy-efficient, easily-installed, easily-maintained coal burning stove.

Parts and Safety

If you buy a stove kit, like the one illustrated in Fig. 9-17, fittings and connections are included in your kit. Chimney parts must be UL listed chimneys, residential type.

You will use a non-combustible floor mat for your stove to stand on. Minimum size floor mats required are shown in Fig. 9-18. Local building codes vary; it's best to check with your local building inspector for sizes and materials.

You need a 6-inch damper to install in the stove-pipe. Be sure that your flue damper is of the type shown in Fig. 9-19 with the small hole in the center. This small hole allows combustion gases a way to escape the stove. It prevents an accidental complete shut-off of your flue and helps prevent burning coals from emitting carbon monoxide gas. If your flue was completely closed this gas would have nowhere to go and could possibly leak into the home, creating a safety hazard.

In addition to the minimum front, side, and rear floor shield clearances you must also shield the floor area below the stovepipe (Fig. 9-20). Place an extra floor shield directly below the stovepipe so that it extends at least 2 inches beyond any point directly below the sides or rear points of the stovepipe. Use a single piece of oversize floor shield that extends at least 2 inches beyond any point directly below the sides or rear stovepipe (Fig. 9-21).

Installation

Place your stove on the non-combustible stove mat so that it is centered on the mat, with the mat extending at least 8 inches on each side and the rear. Also, at least 18 inches of the mat must extend in front of the stove from a point below the stove door. Position your stove at least 36 inches away from any combustible wall on all sides. The stovepipe must

Fig. 9-17. Pudgy Moe cast-iron stove. Courtesy All Nighter Stove Works

be at least 26 inches away from any combustible wall or other material, and at least 24 inches away from any combustible ceiling.

Install a 6-inch damper in the chimney pipe approximately 12 inches from the flue boot on your stove (see Fig. 9-19).

Again, be sure to follow all local building codes as outlined by your local building inspector.

Lid-Lifter and Shaker Grate Tool

Most stove kits come equipped with a combination tool designed to do many jobs (Fig. 9-22). The short end piece is used to lift the lid under the hinged top of the stove. The larger curved end is to be hooked over the coal grate shaker ring on the lower right side of the stove for storage (Fig. 9-23). With it you can operate the shaker grate without fear of being accidentally burned by the hot stove.

The large curved and hooked end also allows you to operate the internal flue without being burned by a hot stove. The large curved end can be hooked over the ash pan handle to put it out for emptying without touching the hot handle.

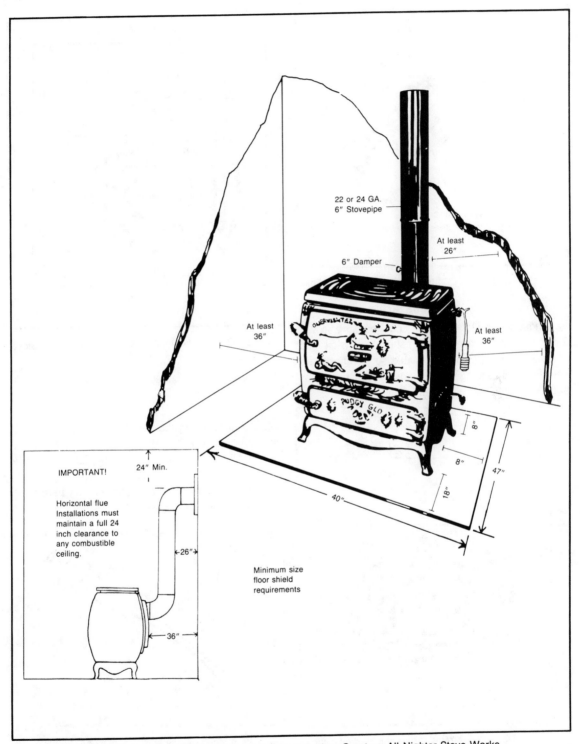

22 or 24 GA.
6" Stovepipe

At least
26"

6" Damper

At least
36"

At least
36"

8"

8"

47"

18"

40"

IMPORTANT!

24" Min.

Horizontal flue
Installations must
maintain a full 24
inch clearance to
any combustible
ceiling.

26"

36"

Minimum size
floor shield
requirements

Fig. 9-18. Layout of free-standing cast-iron stove and floor mat plan. Courtesy All Nighter Stove Works

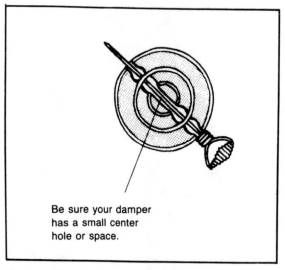

Be sure your damper
has a small center
hole or space.

Fig. 9-19. Flue damper. Courtesy All Nighter Stove Works

Using the Coal Stove

For those who build any type of coal-burning stove, follow the manufacturer's suggestion as to what kind of coal to use. If you are unable to purchase the specific material suggested, contact your local dealer and determine the equivalent.

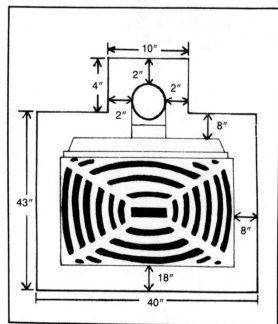

Fig. 9-20. Diagram showing area shield dimensions. Courtesy All Nighter Stove Works

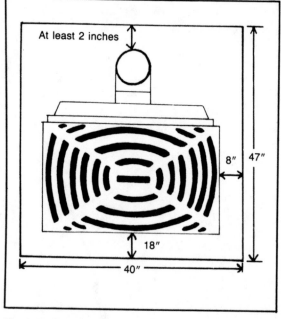

At least 2 inches

Fig. 9-21. Oversize floor shield. Courtesy All Nighter Stove Works

If you are not familiar with building a fire, follow these steps:

■ Place a liberal amount of kindling and paper on top of the grate in the firebox.
■ Open the draft vent and the dampers in the stove and the stovepipe (Fig. 9-24).
■ Light paper and kindling, add dry hardwood when it burns sufficiently.

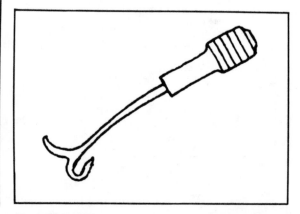

Fig. 9-22. A tool for many jobs. Courtesy All Nighter Stove Works

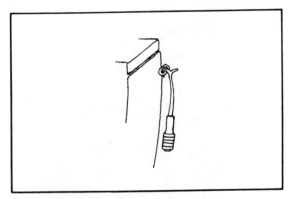

Fig. 9-23. Tool stored on side of stove. Courtesy All Nighter Stove Works

■ With a 4- to 6-inch bed of hardwood embers, add 4 to 6 inches of coal (Fig. 9-25). Allow this to fully ignite and then add additional fuel.

■ When blue flames are produced by the burning of the gases contained in the coal, cut back air controls to get the desired heat output and burning time.

■ Use the handle and side bottom key to shake out the ash as often as necessary. Be gentle.

■ Remove ash from bottom compartment as often as necessary (Fig. 9-26).

To achieve a long and productive burn for the period of time that the stove will be left unattended, you have to "bank" the coal fire. Accomplish this by pushing the heated and burning fuel to the rear and sides of the stove, with the charge gradually tapering down in the front center. Then add a layer of 2 inches or so of new coal (Fig. 9-27).

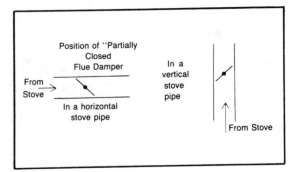

Fig. 9-24. Flue and damper position. Courtesy All Nighter Stove Works

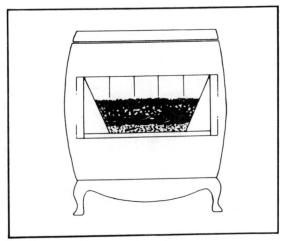

Fig. 9-25. Building a fire from embers of anthracite coal. Courtesy All Nighter Stove Works

To bring the fire back up to temperature at the start of the day, begin with a gentle shaking and open draft intakes and damper fully. Allow to burn ½ to 1 hour. When the red "hot spot" in the center of the firepot is burning sufficiently enough to produce blue flame, add more coal.

Because excess ash will choke a fire, shake the shaker grate. Do not overshake. Shake until a red glow appears in the ash. By doing this, ash will remain on the grate and protect it from the intense heat, thereby allowing the grate a longer life.

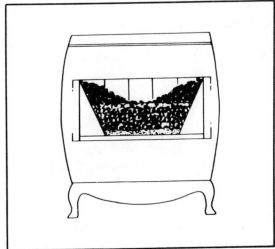

Fig. 9-26. A fully ignited fire. Courtesy All Nighter Stove Works

231

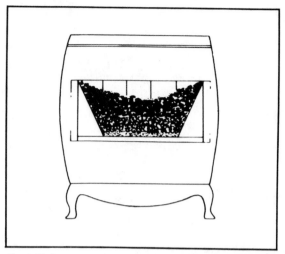

Fig. 9-27. Banking the fire. Courtesy All Nighter Stove Works

Fig. 9-28. Coal grate. Courtesy All Nighter Stove Works

Place ashes in a metal container with a tight-fitting lid. Keep the closed container of ashes on a noncombustible floor or on the ground, well away from all combustible materials. If the ashes are disposed of by burial in soil or otherwise locally dispersed, they should be retained in the closed container until all cinders have thoroughly cooled.

The outside temperature and the temperature you wish to maintain in your home will generally determine the frequency of refueling and of ash removal. Other factors will be the type of fuel available, how well your house is constructed and insulated, the location of the stove in respect to doors, partitions, and open or closed stairways. When cleaning ashes out, leave approximately 1 inch of ash on the coal grate to help retain heat.

Coal Grate

A coal grate is made up of: a base, round shaker grates, a left side, rear retainer, right side, rectangular grate, and two manifold cover plates and a shaker tool (Fig. 9-28).

To install a grate, first clean your stove and stovepipe thoroughly. Put a 6-inch damper in the flue pipe at least 12 inches from the flue boot of the stove. Open the door of the stove and check to see if it has the heat manifold pipe running across its width (Fig. 9-29).

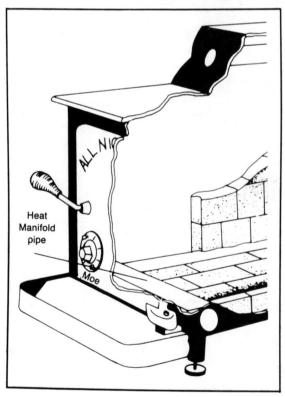

Fig. 9-29. Heat Manifold pipe. Courtesy All Nighter Stove Works

232

If the stove has the manifold pipe then you do not have to use the two manifold cover plates. The base of the grate will simply fit over the manifold pipe. If it does not have the manifold pipe then you have to install the two manifold cover plates over the cutouts in the base of the grate.

Install the rectangular grate with the round shaker grate in place into the stove first. Then insert the rear retainer, right side, and left side on to the grate base. These pieces will lean against the inside walls of your stove.

Remove the cast iron baffle plate from the rear of your stove door. Your coal grate will not work properly unless you remove this air baffle (Fig. 9-30).

MATERIALS TO BURN

Now that you have your fireplace, what should you burn? Or, if you have a coal burning unit, what type of coal should you burn? If you live in an area where coal is readily available, along with wood, you will have a choice. However, availability, cost, and your individual living habits will be the eventual deciding factors.

Wood

Because wood is a renewable fuel supply, it is an attractive substitute for coal, oil, and natural gas for home heating. It is not possible for all of us to heat our homes partially or totally with wood on a continuing basis. In some areas, obtaining heat from

wood is more expensive than producing the same amount of heat from oil or gas. For some people, however, the use of wood might be a suitable alternative to automatic heating systems.

Wood as a fuel has many good qualities. It is much lower in irritating pollutants than other fuels and produces little smoke or soot when properly burned. Wood has a low ash content and the ash that remains from a wood fire can be used as a valuable fertilizer.

A wood fire is easy to start and produces a large quantity of heat in a short time. To dispel a chill quickly a fire is more economical than a large heating system. A wood fire is ideal where heat is required only occasionally. It is excellent for warming the living room on cool days and for supplying extra heat in colder weather.

Not all of the heat produced by the burning of fuel is used because not all of it reaches the space to be heated. The efficiency of a heating system is the ratio of the useful heat to the total amount of heat given off by the fuel. If you hear, for example, that a fireplace is 10 percent efficient, it means that 10 percent of the heat from the fire radiates into the room while 90 percent goes up the chimney.

All conventional heating systems have an efficiency rating. Gas furnaces are usually about 65 percent efficient and oil furnaces are about 60 percent efficient. Air-tight wood stoves have an average efficiency of 50 percent. Other wood-burning devices are less efficient, as indicated in Table 2-9. A range of efficiencies is given for each type of unit because the efficiency of wood-burning units can vary considerably, depending on the rate at which the wood is burned. This rate is determined by the amount of oxygen allowed to reach the fire and by the type of moisture content of the wood. The more oxygen supplied to the fire, the faster, hotter, and brighter it will burn. The drier the wood, the faster it will burn.

The moisture and the particles released when wood burns can condense in the flue and chimney, forming flammable creosote. This can build up on the walls of the flue and could even lead to a flue fire. Flue fires are the greatest danger from heating with wood, and have caused many chimneys to be ruined and even some homes to be burned to the

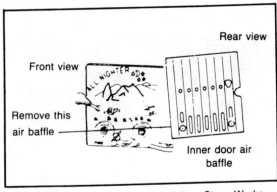

Fig. 9-30. Air baffle. Courtesy All Nighter Stove Works

Table 9-2. Estimated Heating Efficiencies of Wood in Wood-Burning Units.

Built-in fireplaces	0 to 25%
Free Standing fireplaces	
Franklin-type stoves	
Non-airtight stoves	25 to 45%
Airtight stoves	
Airtight Furnaces	45 to 65%

ground. Creosote must be periodically cleaned from chimneys.

Well-seasoned or air-dried wood contains about 15 to 20 percent moisture by total weight. A pound of such wood from most species of trees has about one half the heat value of one pound of good coal. A standard cord of air-dried, dense hardwood has the same heating value as one ton of coal, 200 gallons of domestic flue oil, or 24,000 cubic feet of natural gas. However, a cord of wood will furnish only about one-fourth as much heat to a room as the above amounts of other fuels due to fireplace inefficiency.

Where can you get firewood? The "want ads" in your newspaper is one place to check. Also, contact your state and national forestry office for permission to cut dead, diseased, poorly formed, or weed species of trees. These trees are a hazard to healthy trees and are better removed. Private landowners also are often willing to have waste wood and unhealthy trees removed. Firewood can sometimes be found in dumps and landfills. Industrial wood scraps are often available from sawmills and power companies. Free wood can be obtained from right-of-way clearing, construction sites, or storm damaged areas.

If you're buying firewood, the cord is the standard measure of volume. A cord is 128 cubic feet—a stack of wood 4 feet high, 4 feet deep and 8 feet long. For fireplaces, wood is usually cut into 2-foot lengths. Therefore, firewood measuring 2 feet in length requires a stack of logs 16 feet long and 4 feet high to equal a full standard cord. Portions of a standard cord can be bought, but do not confuse the term cord with "rick," or "pick-up load,"

or a "load" or a "strang" or even a "cord" since these terms often depend on local custom and might contain smaller volumes. For instance, a rick, fireplace, or face cord is a stack of wood 4 feet high by 8 feet long, with a depth equal to whatever length the pieces of wood are cut, usually ⅓ to ½ of a cord (Fig. 9-31).

Firewood is often sold by weight. One ton of air-dried dense hardwood is approximately equal to one-half of a standard cord. If you buy wood by weight, be sure to select the driest wood. Do not pay extra for the water in green wood. Most wood will not burn properly if freshly cut and should be dried or "seasoned." The surest way to have dry wood is to obtain it several months before using it. Splitting the logs will hasten drying time. Smaller round, green logs or green split logs should be stacked outside under a roof for 6 to 10 months before burning. Splitting and stacking wood so that air circulates easily through the stack reduces the drying time.

Most people who gather their own fuel wood burn what is locally available. All wood will burn, and because your main expenditure is your own energy, you can make use of all kinds of wood without much regard to weight or volume. Apple, birch, maple and oak are easy to split; birch and cedar ignite easily; cedar and hemlock produce more sparks than other woods.

Table 9-3 shows the fuel values of various species of wood.

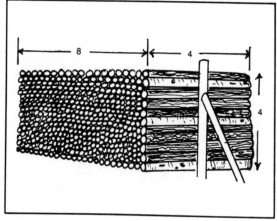

Fig. 9-31. Standard cord of wood. Courtesy USDA

Table 9-3. Fuel Values of Various Species of Wood.

Type of wood	Variety	Pounds per cord	Btu per pound	Equiv. lbs. of coal
Light Woods	Hemlock	1,220	6,410	580
	Pine, white	1,920	6,830	970
	Pine, yellow	2,130	6,660	1,050
	Poplar	2,130	6,660	1,050
	Spruce	1,920	6,830	970
	Willow	1,920	6,830	970
Pine and fir may contain resin that adds to their heating value, inflammability, and smoke-producing properties.				
Medium Weight Woods	Birch	2,880	5,580	1,190
	Chesnut	2,350	5,400	1,260
	Elm	2,350	5,400	940
Wood from garden trees and bushes (lilac, apple, peach, pear) can be included in the medium category.				
Heavy Woods	Ash	3,250	5,450	1,420
	Beech	3,250	5,400	1,300
	Hickory	4,500	5,400	1,800
	Maple, hard	3,310	5,460	1,340
	Oak, white	3,850	5,400	1,540
	Oak, red	3,310	5,460	1,340
	Walnut	3,310	5,460	1,340

Note: Tightly rolled paper logs made of newspapers can be burned effectively in a fireplace. They do not emit sparks and burn slowly. Three or four logs are required to maintain a fire.

Courtesy USDA

Coal

Anthracite coal (hard coal) is generally recommended for most coal-burning stoves. It produces very little smoke, burns with short blue flames, and comes in various sizes: nut (chestnut), pea, and stove coal. Coal sizes are given on Table 9-4.

Coal generally produces 13,500 Btus per pound or 27,100,000 Btus per ton.

Bituminous coal (soft coal) should not be used in stoves because it produces large amounts of smoke and creates coal dust when handled. Bituminous coal also contains high amounts of sulphur.

Cannel coal may be burned in fireplaces but should not be used in a heater or any closed container. Cannel coal contains substantial amounts of volatile matter that can expand and cause small explosions because of the intense heat produced.

Grates are available that allow the stove to burn both wood and coal. The use of such a grate is basically a compromise. No combination fuel stove is as effective as one designed specifically for one fuel only. But the grate is reasonably effective and well worth the compromise, especially for those living in areas where coal is more readily available, thus cheaper than wood. Its main drawback is that it cannot hold the amount of coal that the coal stove can hold so the burn time is shorter. Most coal grate users report loading their coal grate twice daily: once in the morning and once at night. This varies, of course, with the size of stove you purchase. The smaller stove will have a smaller grate.

The major difference between burning wood and coal in a stove is that a coal stove must have its source of primary burning air introduced to the fire from under the grate. Coal consumes twice as much air during the burning process as wood, and this volume of air must pass through the fuel bed completely. If it doesn't, the fire will go out from a lack of oxygen. A good coal stove must be designed to

Table 9-4. Coal Sizes.

Pea Coal	9/16 to 13/16"
Chestnut coal	13/16 to 15/8"
Stove coal	15/8" to 27/16"

provide a strong draft to the underside of the fuel bed.

To derive ultimate efficiency from your coal stove, perhaps one of the most necessary "ingredients" is the patience of the operator. It usually takes a new coal burner much longer to learn the art of burning coal than it takes him to learn wood-burning. Don't get discouraged if your fire goes out on you a few times, just keep trying and experimenting until you get the hang of it. The most common problem is that the novice tends to smother the fire by not providing sufficient primary air or by allowing ash to build up under the firebed.

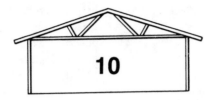

10

Finishing Touches

There are dozens of minor items that make up the finishing touches to your home. Some of the most important: lighting, fans (very popular today for conserving energy) and, of course, painting—inside and out.

LIGHTING

Good lighting properly placed not only makes a home more comfortable and easy to live in, but enhances furnishings, textures, and colors as well.

If you are building a new home or remodeling an older one, the lighting should be a part of the structural and interior plan. Or you can simply start where you are with the lighting you have and improve it. Much can be done at a surprisingly low cost, once you understand and apply the principles of good lighting.

To spot the places in your home where light is needed, make a room-by-room check. Jot down the activities that take place in each room, remembering each member of the family as you do so.

The family living room, for example, is the set-ting for a wide range of activities. Here family members entertain friends and relatives, read, study, telephone, watch television, write letters, or engage in other work or play that require different kinds of lighting. Once you are aware of the places where specific lighting is needed, you can plan the lighting for the activities that take place there. You will find that good lighting greatly increases the usable space within your home.

When planning the lighting of any interior, consider color and finish of walls, ceilings, wood floor or floor coverings, and large drapery areas. These large surfaces reflect and redistribute light within a room. Their lightness or darkness greatly affects the mood of a room.

White surfaces, of course, reflect the greatest amount of available light. Light tints of colors reflect light next best. Somber tones absorb much of the light that falls upon them and reflect little back.

If a large room gets plenty of daylight, you probably can use fairly strong color. Light color on walls make a small room seem larger.

Whatever the room size, try to keep colors within the 35- to 60-percent reflectance range (Table 10-1), floors at least 15 to 35 percent. Matte finishes (flat or low-gloss surfaces) on walls and ceilings diffuse light and reduce reflections of light sources. Glossy, highly-polished or glazed surfaces produce reflected glare.

Light Sources

Incandescent bulbs and fluorescent tubes—in portable and wall lamps, in ceiling and wall fixtures, and in built-in lighting units—are the usual sources of electric light in homes. You must select the right bulb or tube for the purpose you have in mind. Then the bulb or tube has to be placed in an appropriately designed lamp or fixture. And finally, the lamp or fixture must be correctly placed in the room.

Incandescent Bulbs: The incandescent bulb has been the principal home light source for approximately 75 years. It comes in a wide assortment of shapes, colors, sizes, and wattages.

General household bulbs, the most commonly used type, range from 15 to 300 watts. They are available in three finishes: inside frost, inside white (silica-coated), and clear. Inside frost is the older bulb finish still in general use. Use bulbs of this type in well-shielded fixtures. Bulbs with inside white finish (a milky-white coating) are preferred for many home uses. They produce diffused, soft light and help reduce bright spots in thin shielding materials. Decoratively shaped clear bulbs add sparkle to chandeliers or dimmer-controlled simulated candles, and can be a pleasing, low-cost solution to a lighting problem.

Three-way bulbs have two filaments and require three-way sockets. Each filament can be operated separately or in combination. Make sure that a three-way bulb is tightened in the socket so both contacts in the screw-in base are touching firmly. The three lighting levels offered by these bulbs are particularly nice in portable lamps and pull-down fixtures. You can turn the lamp high for reading and sewing, on medium for televiewing, conversation, or entertaining, and on low for a night light or a soft, subdued atmosphere.

Tables 10-2 and 10-3 are guides for incandescent bulbs and sizes and uses for three-way bulbs, respectively.

Table 10-1. Reflectance Range of Various Colors.

Color	Approximate percent reflection	Color	Approximate percent reflection
Whites:		Old gold and pumpkin	34
Dull or flat white	75-90	Rose	29
Light tints:		Deep tones:	
Cream or eggshell	79	Cocoa brown and mauve	24
Ivory	75	Medium green and medium	
		blue	21
Pale pink and pale yellow	75-80	Medium gray	20
Light green, light blue, light			
orchid	70-75	Unsuitably dark colors:	
Soft pink and light peach	69	Dark brown and dark gray	10-15
Light beige or pale gray	70	Olive green	12
Medium tones:		Dark blue, blue-green	5-10
Apricot	56-62	Forest green	7
Pink	64	Natural wood tones:	
Tan, yellow-gold	55	Birch and beech	35-50
Light grays	35-50	Light maple	25-35
Medium turquoise	44	Light oak	25-35
Medium light blue	42	Dark oak and cherry	10-15
Yellow-green	45	Black walnut and mahogany	5-15

Courtesy USDA

Activity	Minimum recommended wattage[1]
Reading, writing, sewing:	
Occasional periods	150.
Prolonged periods	200 or 300.
Grooming:	
Bathroom mirror:	
1 fixture each side of mirror	1-75 or 2-40's.
1 cup-type fixture over mirror	100.
1 fixture over mirror	150.
Bathroom ceiling fixture	150.
Vanity table lamps, in pairs (person seated)	100 each.
Dresser lamps, in pairs (person standing)	150 each.
Kitchen work:	
Ceiling fixture (2 or more in a large area)	150 or 200.
Fixture over sink	150.
Fixture for eating area (separate from workspace)	150.
Shopwork:	
Fixture for workbench (2 or more for long bench)	150.

[1]White bulbs preferred.

Table 10-2. Selection Guide for Incandescent Bulbs.

Courtesy USDA

Table 10-3. Sizes and Uses for Three-Way Bulbs.

Socket and wattage	Description	Where to use
Medium:		
30/70/100	Inside frost or white.	Dressing table or dresser lamps, decorative lamps, small pin up lamps.
50/100/150	Inside frost or white.	End table or small floor and swing-arm lamps.
50/100/150	White or indirect bulb with "built-in" diffusing bowl (R-40).	End table lamps and floor lamps with large, wide harps.
50/200/250	White or frosted bulb.	End table or small floor and swing-arm lamps, study lamps with diffusing bowls.
Mogul (large):		
50/100/150	Inside frost	Small floor and swing-arm lamps and torcheres.
100/200/300	White or frosted bulb.	Table and floor lamps, torcheres.

Courtesy USDA

Dimmer switches are another alternative. They make it possible to light from very low to the maximum output of the bulb.

Tinted bulbs create decorative effects indoors and outdoors. Silica coatings inside these bulbs produce delicate tints of colored light—pink, aqua, yellow, blue, and green. Home uses of these bulbs are best limited to lighting plants, flowers, or art objects. You will need to buy tinted bulbs in higher wattages because they give less light than white bulbs.

Silver-bowl bulbs are standard household bulbs with a silver coating applied to the outside of the rounded end. They direct light upward onto the ceiling or onto a reflector. They come in 60-, 100-, 150-, and 200-watt sizes and are generally used with reflectors in basements, garages, or other work areas.

Reflector bulbs are available with silver coatings either on the inside or outside of the bulbs. The spot-light bulbs direct light in a narrow beam and can accent objects. The floodlight bulbs spread light over a larger area, and are suitable for floodlighting horizontal or vertical surfaces. Typical floodlight sizes include 30-, 50-, 75- and 150-watt. They are also available in tints.

Heat-resistant bulbs, called "par" bulbs because of their parabolic shape, are used outdoors. They are resistant to rain and snow. Common sizes are 75-watt, 150-watt, and up.

Bulbs in decorative shapes are designed to replace bare bulbs in older fixtures and sockets. Some shapes and sizes are made for traditional fixtures (chandeliers and wall scones); others combine contemporary styling and function. Bulb shapes include globe, flame, cone, mushroom, and tubular.

Fluorescent Tubes: Most households use fluorescent lighting in some form. It does offer advantages in home lighting, although know-how is

Table 10-4. Selection Guide for Fluorescent Tubes.

Use	Wattage and color[1]
Reading, writing, sewing:	
Occasional	1 40w or 2 20w, WWX or CWX.
Prolonged	2 40w or 2 30w, WWX or CWX.
Wall lighting (valances, brackets, cornices):	
Small living area (8-foot minimum)	2 40w, WWX or CWX.
Large living area (16-foot minimum)	4 40w, WWX or CWX.
Grooming:	
Bathroom mirror:	
One fixture each side of mirror	2 20w or 2 30w, WWX.
One fixture over mirror	1 40w, WWX or CWX.
Bathroom ceiling fixture	1 40w, WWX.
Luminous ceiling	For 2-foot squares, 4 20w, WWX or CWX
	3-foot squares, 4 30w, WWX or CWX
	4-foot squares, 4 40w, WWX or CWX
	6-foot squares, 6 to 8 40w, WWX or CWX.
Kitchen work:	
Ceiling fixture	2 40w or 2 30w, WWX.
Over sink	2 40w or 2 30w, WWX or CWX.
Counter top lighting	20w or 40w to fill length, WWX.
Dining area (separate from kitchen)	15 or 20 watts for each 30 inches of longest dimension of room area, WWX.
Home workshop	2 40w, CW, CWX, or WWX.

[1]WWX = warm white deluxe; CWX = cool white deluxe; CW = cool white.

Table 10-5. Type of Tube for Ballast Used.

Length of tube (inches)	Wattage Watts	Diameter of tube Inches	Ballast marking[1]
15	14	1½ (T12)	Preheat start.
18	15	1 (T8)	Trigger start or preheat.
		1½ (T12)	Trigger start or preheat.
24	20	1½ (T12)	Preheat or rapid start.
36	30	1 (T8)	Rapid start.
		1½ (T12)	Rapid start or dimming.
48	40	1½ (T12)	Rapid start or dimming.

needed to select and use this light source. For guidance in selecting fluorescent tubes and fixtures, see Tables 10-4 and 10-5.

Because there is a limited selection aesthetically, use for fluorescent lighting in the home is usually in the kitchen, utility room, and bathroom. Fluorescent tubes must be used in fixtures that contain the necessary electrical accessories (Fig. 10-1).

In addition to the long, narrow fluorescent tube, circular fluorescent tubes are also available. They require different fixtures but operate in the same way as the straight tubes. Tube adaptors are also available that screw into a standard incandescent socket, making it possible to use a circline fluorescent bulb without having to install a special fixture. They are available for both ceiling fixtures and table lamps. Initial higher costs for both the adaptor and the bulb are recovered in reduced operating cost and infrequent bulb replacement (Fig. 10-2).

White fluorescent tubes are labeled "standard" and "deluxe." The whiteness of a standard tube is indicated by letters, WW for warm white; CW for cool white. The addition of an "X" to these letters indicates a deluxe tube.

A deluxe warm white (WWX) tube gives a flattering light, can be used with incandescent light, and does not distort colors any more than incandescent light does. A deluxe cool white (CWX) tube simulates daylight and goes nicely with cool color schemes of blue and green. Deluxe tubes are the only fluorescent tubes recommended for home use. They are worth waiting for if your dealer has to order them for you.

Efficient Home Lighting

The efficiency of a light source is measured in lumens of how much light is produced. The amount of energy used is measured in watts. Lumens-per-

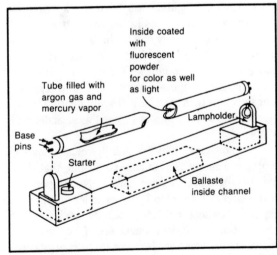

Fig. 10-1. Details of fluorescent bulb fixture. Courtesy USDA

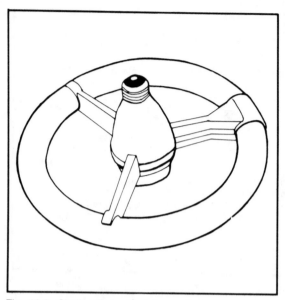

Fig. 10-2. Circline fluorescent lamps fit in incandescent sockets. Courtesy, USDA

Fig. 10-3. Ninety percent of the energy consumed by incandescent bulb is given off as heat. Courtesy USDA

watt ratings are similar to another familiar measure of energy efficiency: miles per gallon. Just as cars have different miles-per-gallon ratings, bulbs (lamps) have different lumens-per-watt ratings. By using energy-efficient bulbs, you can reduce energy use and still have enough light for various activities. The efficiency of different light sources for the home can vary from less than 10 lumens per watt to over 80 lumens per watt.

Fluorescent lighting has been used mainly in commercial environments rather than in the home. This is probably largely due to the fact that it is difficult to find aesthetically pleasing fixtures. However, fluorescent bulbs do not depend on the build-up of heat for light, so the energy wasted as heat is significantly less than with incandescent lighting, making them up to five times as efficient as incandescent lights.

Incandescent bulbs are still the most common source of lighting used in homes, despite that they are also the most energy-wasteful. Ninety percent of the energy consumed by an incandescent bulb is expended as heat, not light. Even though incandescent lighting is inefficient, the choice of the right type and wattage of bulb will help conserve energy (Fig. 10-3).

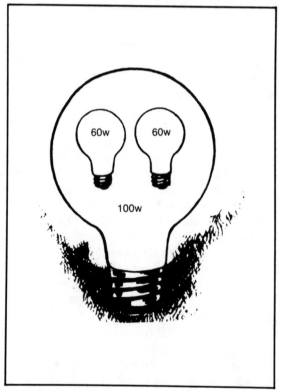

Fig. 10-4. One 100-watt bulb produces more and conserves energy. Courtesy USDA

The lumen output per watt increases as the wattage increases. Therefore, use one large bulb in place of two small bulbs (Fig. 10-4).

One 100-watt bulb at 17 lumens per watt = 1700 lumens.

Two 60-watt bulbs at 13 lumens per watt = 1300 lumens.

For the same amount of wattage and the same use of electricity, the light output is increased by 400 lumens by replacing two 50-watt bulbs with one 100-watt bulb. This not only saves energy but reduces maintenance.

Never use a bulb of a higher wattage than the fixture is designed for, because the fixture could be damaged.

The life span of a bulb, measured in hours, influences the efficiency of an incandescent bulb. Long-life bulbs are the least efficient of the incandescents. They not only cost more, but they also produce 10 to 15 percent fewer lumens per watt. Use long-life bulbs only in places where bulbs are hard to reach and replace.

The light output of a tinted bulb is less than that of a standard bulb with the same wattage.

Various options for making the most of incandescent lighting follows:

Reflector Lamps: Reflector lamps, or R-lamps, offer energy-saving possibilities. Replace a 100-watt bulb in a recessed or directional fixture with a 50-watt R-lamp for specific task lighting or for accent lighting. Instead of trapping light inside, reflector lamps deliver more light where it is needed so that less energy is used. They are available in 25-, 30-, 50-, 75-, and 150-watt sizes. Although a reflector bulb has less lumens per watt than a regular incandescent, more light is actually available, because the light is delivered where it is needed.

Ellipsoidal Reflector Lamps: Also called ER lamps, these are ideal for recessed fixtures, because the beam of light is focused 2 inches ahead of the bulb. This reduces the amount of light and heat that is trapped in the fixture. In a directional fixture, a 75-watt ER bulb delivers more light than a 150-watt reflector bulb.

Parabolic Aluminized Reflector Lamps:

Also called PAR lamps, these have a heavy, durable lens which makes them suitable for outdoor flood and spot lighting. They are long-life bulbs, available in 75-, 150-, and 250-watt sizes.

Guide to Lighting Fixtures

The size of the lighting fixture must be large enough to accommodate the largest watt bulb required to light the area. Often, you need more than one fixture. For example, a large rectangular kitchen may need two 48-inch, two-tube fixtures placed end to end.

Check fixtures carefully before buying them. Here are some points to keep in mind:

■ Incandescent bulbs should be no closer than ¼ inch to enclosing globes or diffusing shields.

■ Top or side ventilation in a fixture keeps temperatures low and extends bulb life.

■ Inside surfaces of shades should be of polished material or finished with white enamel.

■ Shape and dimension of a fixture should help direct light efficiently and uniformly over the area to be lighted.

■ Plain or textured glass or plastic is best for enclosures and shades.

Fixtures in Figs. 10-5 through 10-10 are designs that function well in the house areas specified.

Dimmers. Dimmers add convenience, safety, and flexibility to your home lighting. Gradations of light, from full bright to very dim, are possible simply by turning or pressing a knob.

A low level of lighting is helpful in the care of small children, sick persons, and others who need assistance during the night. House guests, who are unfamiliar with their surroundings, appreciate night lights.

You can make dramatic changes in the mood of a room by softening lights with a dimmer switch. Lights can be lowered when listening to music, enjoying a fire in the hearth, or sitting on an enclosed porch.

Dimmers for incandescent bulbs are simple, compact, and can be mounted in walls in much the same way as off-on switches. Be sure that the

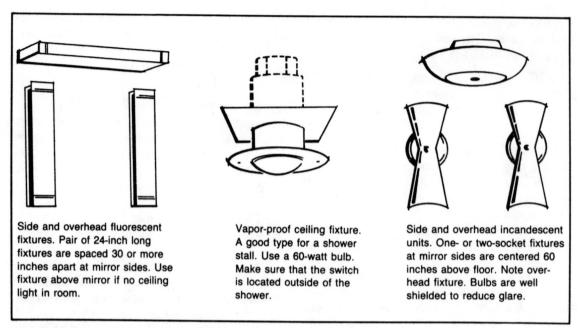

Side and overhead fluorescent fixtures. Pair of 24-inch long fixtures are spaced 30 or more inches apart at mirror sides. Use fixture above mirror if no ceiling light in room.

Vapor-proof ceiling fixture. A good type for a shower stall. Use a 60-watt bulb. Make sure that the switch is located outside of the shower.

Side and overhead incandescent units. One- or two-socket fixtures at mirror sides are centered 60 inches above floor. Note overhead fixture. Bulbs are well shielded to reduce glare.

Fig. 10-5. Bathroom fixtures. Courtesy USDA

wattage capacity of the dimmer control is equal to or more than the total wattage to be controlled. Your local hardware store has them in all necessary wattages and the instructions are plainly marked.

Dimmer-controlled fluorescent fixtures must be preplanned with your power supplier or electric contractor before installation. They are considerably more expensive than incandescent dimmer units. The control combines with a special built-n ballast, and can operate one or more specially-designed fluorescent fixtures as a unit.

Improving Your Lighting

For more brightness and light, finish walls in light

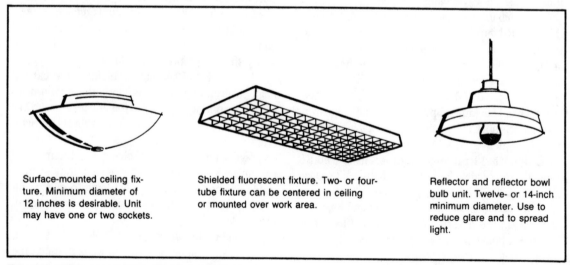

Surface-mounted ceiling fixture. Minimum diameter of 12 inches is desirable. Unit may have one or two sockets.

Shielded fluorescent fixture. Two- or four-tube fixture can be centered in ceiling or mounted over work area.

Reflector and reflector bowl bulb unit. Twelve- or 14-inch minimum diameter. Use to reduce glare and to spread light.

Fig. 10-6. Utility room fixtures. Courtesy USDA

Hanging bowl fixture. Eight-inch diameter. A good choice for lighting a high-ceilinged hall or stairway.

Closed globe fixture. Unit is mounted on ceiling. Choose a white glass globe for diffusion of light.

Wall bracket fixture. May be used to supplement general lighting. Can be mounted on wall near a mirror.

Fig. 10-7. Hallway fixtures. Courtesy USDA

pastel colors and ceilings in white or a pale tint. Flat or low-gloss paint on walls and ceilings helps diffuse light and makes lighting more comfortable. Use sheer curtains or draperies in light or pastel tints.

Add portable lamps for better balance of room lighting.

Install structural lighting (valance and cornice) in living areas where there is only one ceiling light or none. Eight to 20 feet of wall lighting will add a feeling of spaciousness to an average-sized room and make the lighting more flexible.

Replace present bulbs with those of higher wattage, but do not exceed the rated wattage of the fixture. A minimum of 150 watts is needed in many single-socket lamps. For better control of lighting, use three-way bulbs or dimmer switches. If you want a higher-wattage fluorescent unit, the fixture must be changed.

Cover all bare bulbs or tubes in a ceiling fixture with a shade or diffuser. Some of these diffusers clip to the bulb. Others hang from small chains attached to the husk of the fixture. Large diffusers, some-

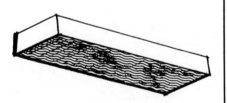

Closed globe unit. Minimum diameter of bowl is 14 inches. White glass gives good diffusion of light.

Shielded fixture. Three or four sockets, 14 to 17-inch diameter. Shallow-wide bowl is desirable.

Fluorescent fixture with diffusing shield. Two or four tubes as needed in a 48-inch unit. For a large kitchen, two 2-tube fixtures can be placed end to end.

Fig. 10-8. Kitchen fixtures. Courtesy USDA

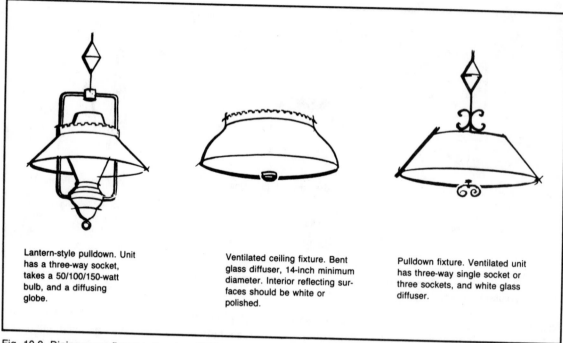

Lantern-style pulldown. Unit has a three-way socket, takes a 50/100/150-watt bulb, and a diffusing globe.

Ventilated ceiling fixture. Bent glass diffuser, 14-inch minimum diameter. Interior reflecting surfaces should be white or polished.

Pulldown fixture. Ventilated unit has three-way single socket or three sockets, and white glass diffuser.

Fig. 10-9. Dining room fixtures. Courtesy USDA

times called adaptors, might have supporting frames that are screwed on the sockets of single-bulb fixtures. An inexpensive way to avoid the glare of bare bulbs in a ceiling fixture is to replace these bulbs with silver bowl bulbs or decorative mushroom-shaped bulbs.

Keep all light sources operating efficiently by replacing blackened bulbs and tubes promptly.

Put in additional convenience outlets if needed for correct lamp placement. Avoid dangerous "cube" plugs and extension cords. Surface wiring strips could be attached along baseboard or counter tops. These strips might be more economical than adding built-in convenience outlets. Be sure any surface wiring system you choose is of the correct size and carries the UL (Underwriters Laboratories) seal

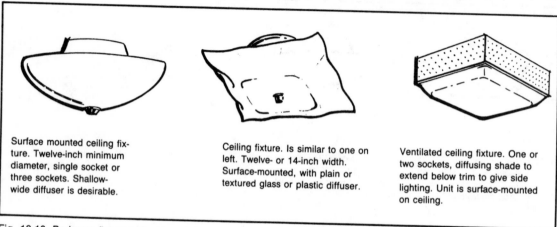

Surface mounted ceiling fixture. Twelve-inch minimum diameter, single socket or three sockets. Shallow-wide diffuser is desirable.

Ceiling fixture. Is similar to one on left. Twelve- or 14-inch width. Surface-mounted, with plain or textured glass or plastic diffuser.

Ventilated ceiling fixture. One or two sockets, diffusing shade to extend below trim to give side lighting. Unit is surface-mounted on ceiling.

Fig. 10-10. Bedroom fixtures. Courtesy USDA

of approval. Ask your electrical power supplier or electrical contractor for the correct size.

Lift the lampshade on a portable lamp with a riser if the bulb is too high. When a bulb is too high it restricts the downward circle of light and shines into the eyes of persons standing near. Risers come in multiples of ½ inch and can be screwed to the top of the harp to lift the shade the amount needed.

Replace the lampshade with a deeper shade if bulb is too low in lamp and bulb shows beneath lower edge of shade. Or, if you prefer, use a shorter harp or a different diffusing bowl.

If a lamp base is too short, set the base on wood, marble, ceramic, or metal blocks to raise lamp to proper height. For ease in handling, cement the block to the base.

Get replacement shades for table lamps if present shades do not meet specifications. Choose shades made of translucent materials with white linings and open tops.

Lighting Maintenance

Home lighting equipment needs regular care and cleaning to keep it operating efficiently. A collection of dirt and dust on bulbs, tubes, diffusion bowls, lampshades, and fixtures can cause a substantial loss in light output. It's a good idea to clean all lighting equipment at least four to six times a year; bowl-type portable lamps should be cleaned monthly. Better still, get into the habit of dusting them when you are doing your regular housework.

Here are some suggestions for taking care of lamps and electrical parts: Wash glass and plastic diffusers and shields in a detergent solution, rinse in clear warm water, and dry. Wipe bulbs and tubes with a damp, soapy cloth, and dry well. Dust wood and metal lamp bases with a soft cloth and apply a thin coat of wax. Glass, pottery, marble, chrome, and onyx bases can be washed with a damp soapy cloth, dried, and waxed.

Lampshades may be cleaned by using a vacuum cleaner with a soft brush attachment, or dry-cleaned. Silk or rayon shades that are hand-sewn to the frame and have no glued trimmings, may be washed in mild, lukewarm suds, and rinsed in clear water. Dry shades quickly to prevent rusting of frames. Wipe parchment shades with a dry cloth.

Remove plastic wrappings from lampshades before using. Wrappings create glare and could warp the frame and wrinkle the shade fabric. Some are fire hazards.

Replace all darkened bulbs. A darkened bulb can reduce light output 25 to 50 percent, while still using almost the same amount of current as a new bulb operating at correct wattage. Darkened bulbs may be used in closets or hallways where less light is needed.

Replace fluorescent tubes that flicker and any tubes that have darkened ends. A long delay in starting indicates a new starter is probably needed. If a humming sound develops in a fluorescent fixture, the ballast might need to be remounted or replaced.

FANS

In the past few years energy efficiency has been foremost in our minds as we build our new homes or remodel older homes. For this reason fans have made a comeback. They are used primarily to conserve energy, but they also add to the decor. They are easy to install and come in a variety of designs.

Figure 10-11 illustrates an ideal fan for living and dining rooms, and bedrooms. It can be purchased with a light attachment, as can most fans. The fan in Fig. 10-12 is ideal for use in a bathroom, hallway, or closed-in porch. In Fig. 10-13 is a fan perfect for a dining area. The choices are varied and only the individual can decide which is best for their particular area.

Most fans require the same type of assembly and installation. Be sure to read the manufacturer's instructions carefully before you begin. If you are not able to do the minor electrical work, it's best that you consult a qualified electrician.

Installation

Check the parts in the packing box from the manufacturer before starting installation (Fig. 10-14). Most fans consist of:

—1 fan unit
—4 blades

Fig. 10-11. A fan suitable for many locations. Murray Feiss Imports, New York

—1 canopy
—1 mounting bracket
—1 down rod
—1 screw pack

Caution: Attach blades *after* motor housing is hung and is in place. Keep fan motor housing in carton until it is ready to be installed—laying motor housing on its side could cause decorative casing to shift.

Be sure power is off to supply circuit intended for connections.

Fig. 10-12. The Sebring. Courtesy Leslie-Locke, A Questor Company

Fig. 10-13. Ideal fan for dining area. Courtesy Leslie-Locke, a Questor Company

Slide the canopy up the down rod with the dish facing the hanger.

Pass the wires from the top of the fan through the down rod. Thread the rod—with the canopy in place—into the top of the fan. Tighten. Tighten the lock screw securely. *Caution:* Make sure that the set screw on the fan motor is tightened through the hole in the threaded portion of the down rod when installing fan.

Attach the mounting bracket to the junction box in the ceiling. Be sure the box is secured in the ceiling. You must have a 2-by-4-inch piece of wood or other approved hanger mounted in the attic, supporting the fan and junction box. The mounting bracket must have the safety notches facing toward the floor. The junction box must be able to support 40 pounds.

Hang the fan by down rod on the mounting bracket attached to the junction box in ceiling. Be sure power is off. Connect the white wire from the fan to the white wire in the ceiling. Next, connect the black wire from the fan to the black wire in the ceiling.

Attach the green ground wire to "J" box body.

Slide the canopy up within ⅛ inch of the ceiling and set the screws.

Attach the blade holders with blades to the motor with the screws in screw pack. Attach one blade to the motor. Then turn the motor by hand until the blade is opposite of you. Attach the second blade. Repeat for the third and fourth blade.

Turn on the power. The pull chain will control the on/off function, and the black knob on the side of the switch housing will set the desired speed. (In some fans, this black knob would be white, it depends upon the manufacturer.)

Lamp Option

If lamp option is used, remove switch housing cap and remove the ⅛-inch plug from cap. Unscrew and remove the lower half of the lamp socket in lamp assembly.

Pass the red and white wires marked "to light" through the hole in the switch housing cap and through the stem in the lamp assembly. Connect red wire to the brass-colored terminal on the lamp socket and white wire to the silver-colored terminal.

Replace switch housing cap on switch housing.

Carefully thread stem on the lamp assembly while feeding extra wire length into switch housing, and tighten lamp assembly. Replace the lower half of the lamp socket and install a light bulb (60-watt maximum).

Figures 10-15 through 10-19 are diagrams of the fan parts and installation.

Troubleshooting

Your fan might make a slight noise on low speed. Do not be concerned, this is an electrical noise caused by reduction of watts going through the condenser.

If the fan does not start, check wiring to insure that all connections are secured. Check capacitor and solid-state speed control. If parts are not allowing a good current flow, replacement parts will be provided by the manufacturer.

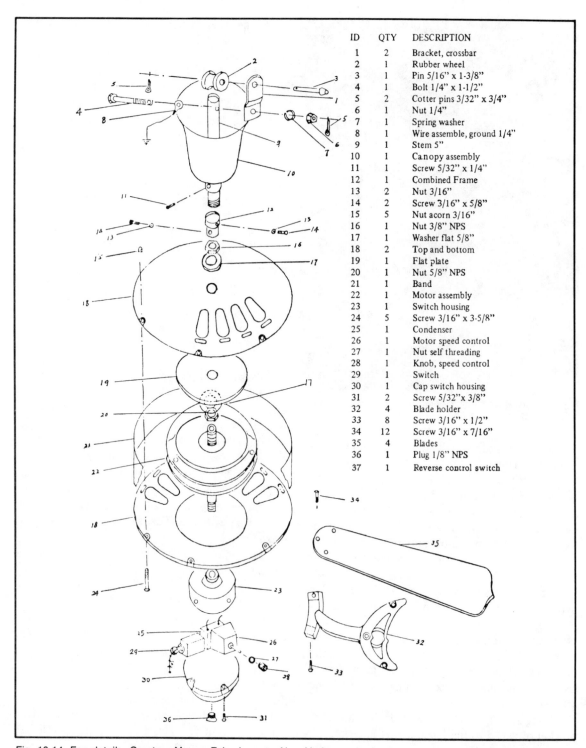

ID	QTY	DESCRIPTION
1	2	Bracket, crossbar
2	1	Rubber wheel
3	1	Pin 5/16" x 1-3/8"
4	1	Bolt 1/4" x 1-1/2"
5	2	Cotter pins 3/32" x 3/4"
6	1	Nut 1/4"
7	1	Spring washer
8	1	Wire assemble, ground 1/4"
9	1	Stem 5"
10	1	Canopy assembly
11	1	Screw 5/32" x 1/4"
12	1	Combined Frame
13	2	Nut 3/16"
14	2	Screw 3/16" x 5/8"
15	5	Nut acorn 3/16"
16	1	Nut 3/8" NPS
17	1	Washer flat 5/8"
18	2	Top and bottom
19	1	Flat plate
20	1	Nut 5/8" NPS
21	1	Band
22	1	Motor assembly
23	1	Switch housing
24	5	Screw 3/16" x 3-5/8"
25	1	Condenser
26	1	Motor speed control
27	1	Nut self threading
28	1	Knob, speed control
29	1	Switch
30	1	Cap switch housing
31	2	Screw 5/32" x 3/8"
32	4	Blade holder
33	8	Screw 3/16" x 1/2"
34	12	Screw 3/16" x 7/16"
35	4	Blades
36	1	Plug 1/8" NPS
37	1	Reverse control switch

Fig. 10-14. Fan details. Courtesy Murray Feiss Imports, New York

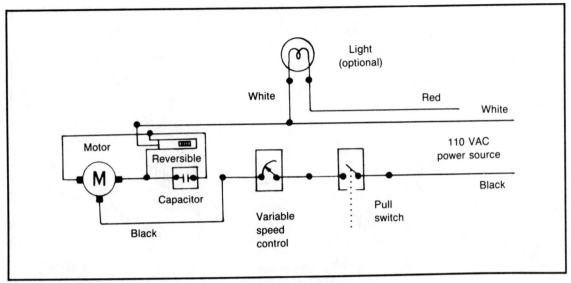

Fig. 10-15. Wiring diagram of fan. Courtesy Murray Feiss Imports, New York

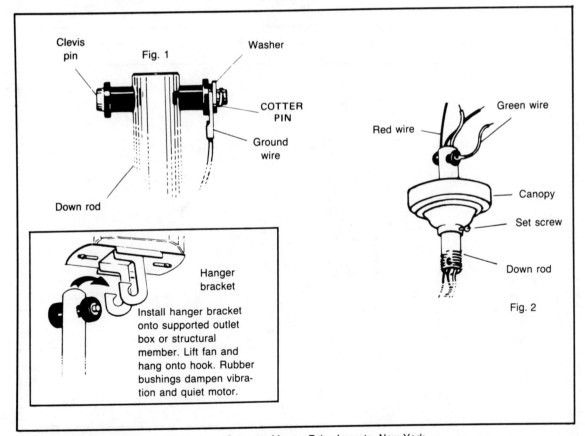

Fig. 10-16. Fan detail, wiring and brackets. Courtesy Murray Feiss Imports, New York

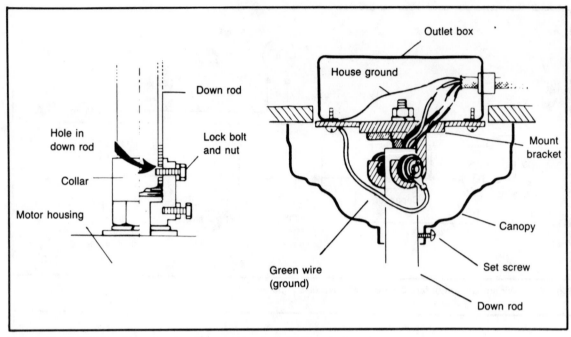

Fig. 10-17. Fan detail, wiring. Courtesy Murray Feiss Imports, New York

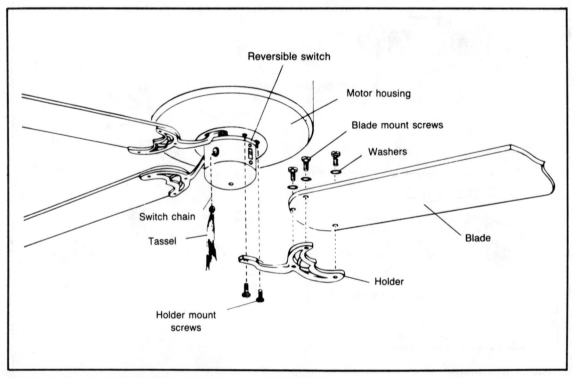

Fig. 10-18. Fan detail. Courtesy Murray Feiss Imports, New York

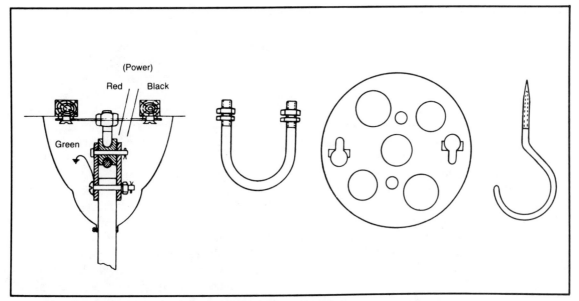

Fig. 10-19. Wire connection. Courtesy Murray Feiss Imports, New York

If the fan wobbles 1 inch or more, you might have uneven blades, or arm angles or warped blades due to shipping or damage prior to installation.

Check angles of blades. Let the fan run for a few hours. If wobble is not corrected, "switch" two blades that are next to each other.

If the motor would happen to jam or freeze, the fan motor would need replacing by the manufacturer.

PAINTING INSIDE AND OUT

If yours is a new house not yet painted, you have a wide range of finishes to choose from. If the siding on your house is knotty, it might be best finished with a penetrating stain rather than paint. If there is only an occasional knot in your siding, however, paint would work well. In this case, scrape the knots and wipe with a cloth that has been wetted with paint thinner. Then apply a good knot sealer, or an extra prime coat over the knots.

Hardboard siding is available smooth and textured, and is generally sold preprimed. The factory-applied primer is compatible with a wide variety of paints and stains.

Textured plywood sidings are available in various species and many patterns. They are best finished with high-quality exterior stains because the appearance is compatible with the rustic siding. Textured plywood may also be painted, using top-quality acrylic latex exterior house paint with companion stain-resistant primer.

With either stain or paint, apply at least the first coat with a brush to mechanically work the finish well into the textured surface. For brushed or abraded plywood, sometimes called relief-grain plywood, follow the finishing recommendations of the specific panel manufacturer, or use a semi-transparent penetrating stain. Where check-free plywood siding is desired, use medium density overlaid plywood.

Medium density overlaid plywood sidings are available with both smooth and textured or patterned surfaces. The overlay provides a stable base and is designed primarily for paint finishes. Opaque stains may also be used. The service life of finishes on medium density overlaid plywood is significantly greater than on non-overlaid plywood.

Natural Finishes

Natural finishes help to retain or enhance the natural color and grain of wood. Such finishes are most extensively used on the attractive siding woods such as redwood, western red cedar, and Philippine mahogany, but are not limited to these uses. They are

253

relatively easy to apply and economical to maintain. The natural finish family includes water-repellent preservatives, bleaches, and penetrating or semi-transparent stains. All are either unpigmented (clear) or pigmented very slightly. Hence, "natural" finishes.

Paintable Water-Repellent Preservatives (WRP): Paintable water-repellent preservatives (not silicone type) repel liquids such as rainwater and dew, and thereby reduce the swelling and shrinking of wood. They also protect wood against mildew and decay. Paintable water-repellent preservatives are easy to apply by brush, roller, or by dripping. They penetrate into wood, leaving its appearance relatively unchanged except for a slight darkening. However, treatment with WRP slows down the weathering process and protects against water staining, as well as against mildew attack.

Where paintable water-repellent preservative is the sole treatment to be applied to exterior wood surfaces, two coats are recommended. Best results are obtained when the first coat is applied to the back, face edges, and ends of the wood before it is nailed into place. After installation the second coat should be brushed over all exposed wood surfaces. As weathering progress, the color of WRP-treated wood might lighten.

The frequency with which the water-repellent preservative treatment needs to be renewed is dependent upon climatic conditions. In relatively dry areas, the treatment retains its effectiveness longer than in areas subject to extensive rainfall. In dry areas treatment might not need to be renewed for three to five years. In wet areas, renewal of the treatment might be required after 12 to 18 months. Successive retreatment may be extended to 2 years or more. Darkening of the wood or blotchy discoloration are indications that the treatment has lost its effectiveness. This can be verified by splashing one-half cup of water against the treated wood surface. If the water balls up and runs off the surface, the treatment is still effective. If the water soaks quickly into the wood, it is time to refinish. A single recoat should suffice.

Water-repellent preservative is also an excellent pretreatment for wood that is to be painted. Exterior millwork (windows, doors, and trim) that is simply primed and painted could eventually show signs of paint failure around the joints. Rainwater or dew can seep into the joints and swell the wood and, upon drying out, crack the paint film over the joints. Such paint failure can be significantly reduced by brushing water-repellent preservative over the wood surface and into the joints prior to priming. The label should indicate that the WRP is paintable. Allow a minimum of 72 hours for the WRP to dry before painting or staining.

Bleaches: Bleaches are recommended *only* for solid wood siding. Homeowners who want their houses to have a weathered appearance quickly may achieve it by applying a bleach or bleaching oil. Bleaching oils are available in many paint stores. In addition to a bleaching chemical, the better bleaching oils contain pigments to impart a grayed appearance to wood, and an agent to protect the finish against mildew.

By means of chemical reaction with the wood, bleaches hasten the natural color changes brought on by weathering and eliminate the darkening that often occurs when wood weathers naturally. On new wood, one or two coats of bleach (generally two) are recommended. The original application is often the only one necessary. Reapplication of the bleach is required only if the wood begins to darken or if the bleaching becomes uneven.

Because the bleaching action is aided by moisture, together with sunlight, it is helpful to periodically spray bleached surfaces with water from the garden hose.

Semi-Transparent Stains: Semi-transparent stains, sometimes called penetrating stains, contain a small amount of pigment which allows them to alter the natural color of wood, but only partially obscure the grain or texture. They are generally offered in natural, wood-tone colors, and should be of the oil-base type.

Two coats of penetrating stain are generally recommended on new wood, and application is best done by brush or flat applicator. Roller or spray application, followed by brushing, may be used on smooth wood and textured surfaces. Take care on windy, dry days to avoid lap marks.

Penetrating stains leave a flat or dull finish. They are a "breathing" type of finish since they do not form a continuous film or coating on the surface of the wood.

A penetrating stain finish is gradually worn away by the weather. When the erosion progresses to the point that portions of the wood show through, it is time to refinish. A single coat is generally all that is required.

Solid-Color (Opaque) Stains

The increasing use of textured wood siding, an ideal surface for stains, has added to the popularity of solid-color stains. Also called heavy-bodied stains, they are made with a much higher concentration of pigment than penetrating stains. As a result, solid-color stains have a higher hiding power; they obscure the natural color and grain of wood. They are more like paint than stain.

The solid-color stains are available in nearly as wide a variety of colors as paint. They are available in water-thinned (latex) as well as in solvent-thinned types. The major advantages of the water-thinned stains include ease of cleanup, slower finish erosion, and better color retention. However, water-thinned stains are more difficult to apply by brush without showing lap marks.

As a rule, only a single coat of solid-color stain is applied, but two coats will provide better and longer service. Any of the conventional methods of application may be used to apply the stain to smooth wood surfaces, but brush application is best.

Transparent Coatings

Varnishes, synthetic resins, and other clear film finishes are not generally recommended for exterior use on wood products exposed to direct sunlight. Such coatings permit ultraviolet light to degrade the wood surface, causing loss of adhesion, as well as cracking and peeling of the finish in less than two years. This condition occurs no matter how many coats are applied. Eventually the film will have to be completely removed by sanding or with a varnish remover.

Exterior marine or spar varnishes can be used effectively on exterior doors that are adequately protected from rain and sun.

Although oil treatments, such as boiled linseed oil, can produce a dark, pleasing effect on new exterior wood surfaces, they are subject to all the problems of a varnish as described above. In addition, mildew growth and dirt accumulation can be severe on oil-treated wood.

Paints

Of the various types of finishes available for use on exterior wood surfaces, properly applied paints offer the widest selection of colors and provide the most protection against weathering of the wood. Paints usually give the longest service life or durability possible.

Because the first coat of paint on new wood is the most important, follow the manufacturer's recommended system. Its selection is determined by the type and brand of topcoat paint you wish to use. Some paint manufacturers offer a primer that is specially designed for use with each of the various types of topcoat paints they make. Other manufacturers offer a "universal primer" that may be used under a variety of topcoat paints. The primer-topcoat system should always be made by the same manufacturer.

Paint primers may be either alkyd resin base (solvent-thinned) or latex resin base (water-thinned). Solvent-thinned primers have been especially recommended for woods such as redwood and western red cedar. They seal the wood against moisture and protect against extractive staining. Latex primers have excellent flexibility and are preferable for use on Douglas fir and southern pine. They should, however, be a stain-resistant type.

Most paint manufacturers print recommended spreading rates for their primer on the label. These should be followed. If such a spreading rate is not available, a good rule to follow is to apply the primer in sufficient thickness to obscure the grain of the wood. This will take one gallon of the primer to cover between 400 and 450 square feet of a smooth surface. Textured surfaces can require up to twice as much paint.

Topcoat Paints: Topcoat paints may be classified into two basic categories: solvent-thinned paints and water-thinned paints.

Solvent-thinned paints (the solvent is usually mineral spirits) are generally alkyd resin-base or oil-base paints. Such paints have been available for many years.

The water-thinned paints are emulsion paints or, as they are generally called, "latex paints." They are made of acrylic or polyvinyl cetate resins or other emulsion resins. Latex paints dry rapidly, permitting the finish to be recoated the same day.

Latex paints are gaining popularity, especially with homeowners who do their own painting. Reasons for this are obvious: latex paints are easy to apply and equipment is easy to clean. Latex paints are also less likely to blister, and have excellent durability and color retention. However, a coat of latex paint, when dry, is generally thinner than a similar coat of oil-base paint.

If only the second or upper story of your house is sided with wood siding and the first or lower story is brick or other masonry, you will want to be assured that the paint on the wood siding does not chalk excessively. Otherwise, it will be washed (by rainwater or dew) down over the masonry and discolor it. Tell your paint dealer of the situation so he can provide you with proper paint products.

Application of Paint

Now that you have made the important choice of coating, you would be wise to stand back and consider exactly how you will go about applying it to your house. Painting a house, whether or not it has been painted before, takes time and effort. So it is very important to do it right the first time. Planning ahead pays off. If you have never painted a house before, the whole project probably seems very complex. It doesn't have to be. By following a few basic guidelines and approaching the job one step at a time, your house can look as if painted by a professional.

For the original paint job on new wood surfaces, three coats are recommended. A three-coat application will perform better and last much longer than a two-coat. However, most original paint jobs are limited to two coats. If you plan to limit yours to two coats, be certain to apply both generously, toward

the lower spreading rate of the range specified by the manufacturer on the label of the container. Again, manufacturer's instructions for use of primer and topcoat should be followed. On factory-primed sidings, the factory primer will take the place of one of the required coats.

Repaint work is best limited to a single coat in a color similar to the original color. If you plan a color change, apply two coats. Shake-and-shingle paints and the flat oil-base paints are not recommended because of their short service life.

Finish coats of house paint can normally be applied at a spreading rate of about 500 square feet of surface per gallon on smooth surfaces. The primer can be applied at about 450 square feet per gallon. Rough or textured surfaces might require up to twice as much paint. Check the container label for recommended spreading rates because they can vary.

With these figures and the house dimensions, the approximate number of gallons of paint and primer required can easily be determined using this procedure:

■ Average height of the house equals the distance from foundation to eaves for flat roof types; add 2 feet for pitched roofs.

■ Average height multiplied by the distance around the foundation equals the surface area in square feet.

■ The surface area divided by 450 equals number of gallons of primer required.

■ Surface area divided by 500 equals number of gallons of finish paint required for each coat.

Brush application is the preferred method for applying the primer, regardless of the type of substrate. It is essential on textured surfaces, whether mechanically textured or roughened by weathering. Topcoats of paint can be satisfactorily applied by brush, roller, or pad. If a roller is used to apply topcoats to textured surfaces, follow up by brushing the paint into the rough surface. This gives a good appearance. Spray painting is the most economical on large, uninterrupted expanses, but is best left to the professionals.

New wood should be painted promptly (within two weeks) after its installation. If you find that this cannot be done, protect the bare wood as soon as possible from rain and heavy dew and mildew by brushing a paintable water-repellent preservative solution on the siding trim, and into all joints. Wood so treated should be allowed to dry for a few days prior to painting or staining.

The best time to paint is during clear, dry weather. Temperatures must be above 50 degrees F. Latex paints may be applied even though the surface to be painted is damp from condensation or rain. Solvent-thinned paints should be applied only to a dry surface.

If the outside temperature is high (70 degrees F or higher), it is best to paint those surfaces already reached by shade. This is known as "following the sun around the house." To avoid the wrinkling and flatting of solvent-thinned paints, and water marks on latex-painted surfaces, do not paint late in the day in early spring or late fall when heavy dew is likely to occur. Here is another reason for avoiding the application of latex paints in the afternoon in the late fall: A sudden drop in the temperature is likely to occur at night, which can result in failure of latex paint to form a proper film.

Painting New Wood

The first step in painting new wood is to prepare the surface properly. Not a great deal of preparatory work is required on the exterior of a newly constructed house. Dry-brush all foreign matter such as mud flecks, dust, dirt, and loose wood particles from the surface to be painted. As stated previously, scrape knots and wipe them with a cloth wetted with paint thinner, and then seal or spot-prime them.

Caulk cracks and open joints in the siding with a paintable polysulfide, latex, acrylic, or butyl caulk.

All edges of lumber, plywood, and hardboard should be primed before installation. Only stain-resistant nails (hot-dipped galvanized aluminum or stainless steel) should be used to fasten wood siding. If other nails are used, rust marks that show through the paint are likely to be a never-ending problem. These stains are the result of a chemical reaction between the iron in the nails and the natural chemicals found in wood.

Follow the manufacturer's directions with re-

spect to the spreading rate at which the primer should be applied and the time it should be permitted to dry before the topcoat is applied. However, no longer than a week should elapse between the application of the primer and the first coat of topcoat paint.

As stated previously, a three-coat job on new wood surfaces provides better and longer-lasting service. However, if three coats cannot be applied to the entire exterior of the house, three coats should at least be applied to the south and west sides of the house, since these sides usually receive the most severe weathering. Three coats should also be applied to exposed horizontal or near-horizontal surfaces such as window sills, hand rails, etc., because paint is subjected to much greater weathering on these surfaces than it is on vertical surfaces. The one exception to this rule involves outside decks, which are better finished with a penetrating stain than with a paint.

Painting Metal Surfaces

Generally, conventional finish coats are not designed for use on unprimed metal. Always use a recommended metal primer prior to applying finish coats.

On rusted iron surfaces, first remove rust, then prime the surface with a coat of zinc chromate metal primer. For galvanized surfaces, a prime coating containing zinc dust is recommended. Latex house paints have also been found to be satisfactory for priming galvanized and aluminum surfaces.

Repainting

Much repainting today is done not out of need, but out of the desire to simply change the color of the exterior of the house. To assure good adhesion and lasting performance, the paint on the house should have weathered sufficiently before being repainted. If it hasn't, it is advisable to scuff sand (sand lightly) those surfaces that are relatively protected from the weather, such as the north side of the house. Wash soffits, and the surfaces immediately under roof overhangs, with plain water, using a sponge or cloth, and rinse these areas thoroughly before repainting.

A better repainting schedule is one dictated by the amount of wear or erosion the paint has undergone. Repainting should be undertaken when the old paint has eroded to the point where it no longer covers the wood completely. If the paint has completely worn away, leaving bare wood exposed, repaint promptly, after having prepared the surface. Preparation includes sanding the bare wood and surrounding area with fine- to medium-grit sand paper and removing the dust with a cloth or dry brush.

Where paint is peeling and bare wood is exposed, remove all loose and curled paint by scraping and sanding. Sand the exposed wood and the paint around the bare areas, and take care to smooth the edges of the sound paint film. Sand rust spots over and around nail heads to remove the rust, and apply a rust-inhibiting primer to seal the nail heads. Prime bare wood areas in accordance with manufacturer's directions, feathering the primer beyond the bare wood and out onto the sound painted area.

If the old paint is only chalking, faded, and dirty, the preparatory work might be limited to scrubbing the surface with a household detergent or trisodium phosphate mixed in water. Following the scrubbing, rinse thoroughly with clean water and allow the surface to dry before applying paint.

Caulk splits and open joints with a paintable polysulfide, acrylic, or butyl caulk. After the caulk has been smoothed in place, apply primer over the caulked area, feathering it out onto the adjacent sound paint film. Part of the surface preparatory work should consist of inspecting the exterior of the house for possible avenues of moisture entry.

Spotty patches that look like dirt but do not come off when scrubbed with a detergent solution are probably mildew—which is prevalent in warm, humid climates. Mildew can occur on both sunny and shaded sides of buildings, but is more likely to grow on the shaded side, particularly behind trees or shrubs where air movement is restricted.

Mildew must be killed and removed before repainting, or it will simply come through the newly applied coat of paint. To kill the mildew and clean the surface for repainting, scrub the affected areas with the following solution: ⅔ cup trisodium phosphate, 1 quart household bleach (do not mix bleach with ammonia or any detergents or cleaners containing ammonia), 3 quarts warm water (enough to make

1 gallon of solution).

It is advisable to wear rubber gloves and goggles when applying this solution. Scrub with a fairly soft brush. When clean, rinse thoroughly with fresh water. Avoid splashing the solution on shrubbery or glass.

If mildew is a serious problem in your area, apply only paint (and primer) that contains a mildewcide. If not, the mildew problem will be a recurrent one.

Moisture Control

Paint does not fail without cause. Blistering, peeling, and flaking, as well as extractive staining of the paint (on such woods as redwood and western red cedar), are caused by moisture—as are most other forms of paint failure.

Moisture accumulation in the sidewalls of a house is caused by moisture vapor that originates within the house and permeates the walls, only to condense as water. It also occurs when outside water (rainwater or dew) enters the sidewalls through joints or other openings in the siding. Whatever the cause, it is likely to result in failure of the exterior paint in one form or another. Unless the source of moisture is located and eliminated, even more serious trouble may develop, such as decay of the studs, plates, and sheathing. Fortunately, there are proven methods for preventing the entry of moisture into the sidewalls of the house.

Interior Moisture: Moisture vapor is generated within every house in the form of steam from cooking, and hot water from the shower and from humidifiers. Modern appliances such as dishwashers, and clothes washers and dryers all add moisture to home interiors.

Part of the task of insulating should include installing a vapor barrier on the warm side of the exterior walls when the house is constructed. If interior moisture is causing paint on the outside of your house to fail, make every effort to reduce the relative humidity in the house. Exhaust fans in the kitchen, bath, and laundry will carry much of the moisture in these rooms outside. Attics, too, should be well-ventilated, especially during the winter months. For gable roofs, use screened louvres (¼-inch mesh).

If your house was built with a crawlspace, ground moisture might be contributing to excessively high humidity in the house. Check to determine if a vapor barrier ground cover was installed over the ground. Polyethylene film 4 mils thick serves as an excellent ground cover. It should be lapped about 6 inches at the joints, free of punctures, and snugged up well around pipes. Check to determine if the area is properly drained and free of pools. Also, see that the crawlspace vents are open.

Exterior Moisture: Outside moisture can be prevented from gaining entry into the sidewalls by:

■ Installing proper flashing where it is lacking.
■ Repairing or replacing rusted or damaged flashing.
■ Caulking splits and open joints.
■ Preventing winter ice dams that might form at the house eaves.
■ Applying water-repellent preservative at all joints, lap and butt.

Paint failures caused by moisture in the sidewalls of the house will continue to occur unless the source of moisture is located and eliminated.

Interior Finishes

The type of finish depends largely upon type of area and the use to which the area will be put. The various interior areas and finish systems employed in each are summarized in Table 10-6. Wood surfaces can be finished either with a clear finish or paint. Plaster-base materials are painted.

Wood Floors: Hardwood floors of oak, birch, beech, and maple are usually finished by applying two coats of wood seal, also called floor seal, with light sanding between coats. A final coat of paste wax is then applied and buffed. This finish is easily maintained by rewaxing. The final coat can also be a varnish instead of a sealer. The varnish finishes are used when a high gloss is desired.

When floors are to be painted, an undercoater is used, and then at least one topcoat of floor and deck enamel is applied.

Wood Paneling and Trim: Wood trim and paneling are most commonly finished with a clear wood sealer or a stain-sealer combination, then topcoated after sanding with at least one additional coat of sealer or varnish. The final coat of sealer or varnish can also be covered with a heavy coat of paste wax to produce a surface that is easily maintained by rewaxing. Good depth in a clear finish can be achieved by finishing first with one coat of a high-gloss varnish followed with a final coat of semi-gloss varnish.

Wood trim of a nonporous species such as pine can also be painted by first applying a coat of primer or undercoater, followed with a coat of latex, flat or semi-gloss oil-base paint. Semi-gloss and gloss paints

Table 10-6. Guide for Interior Finishes.

	Primer or under- coater	Rub- ber latex	Flat oil paint	Semi- gloss paint	Floor (wood) seal[1]	Var- nish[1]	Floor or deck enamel
Wood floors	X				X	X	X
Wood paneling and trim	X	X	X	X	X	X	
Kitchen and bathroom walls	X			X			
Dry wall and plaster	X	X	X				

[1]Paste wax can be applied over floor seal and varnish base.

are more resistant to soiling and more easily cleaned by washing than the flat oil and latex paints. Trim of porous wood species such as oak and mahogany requires filling before painting.

Kitchen and Bathroom Walls: Kitchen and bathroom walls, which normally are plaster or drywall construction, are finished best with a coat of undercoater and two coats of semi-gloss enamel. This type of finish wears well, is easy to clean, and is quite resistant to moisture.

Drywall and Plaster: Plaster and drywall surfaces, which account for the major portion of the interior area, are finished with two coats of either flat oil or latex paint. An initial treatment with size or sealer will improve holdout (reduce penetration of succeeding coats) and thus reduce the quantity of paint required for good coverage.

Additional Tips

If you have not previously painted a house, start on the back or a side of the house, whichever is the more inconspicuous. By the time you get around to the front of the house you will have gained experience and confidence.

Painting an entire house takes time and, although it is necessary to do the original paint job all at one time, there is no reason why you should not spread the repaint job over several years. Paint weathers fastest on the south side of the house, next fastest on the west side, then on the east side, and slowest on the north side. Accordingly, the south side of your house will require repainting first—approximately five to seven years after the original paint job.

A year or two later, dependent on the condition of the old paint, you might repaint the west side. This would be followed by the east side and, unless you change to a different color, the north side might simply be washed down. In this manner, you can spread the task of repainting over a period of several years.

Start painting at the highest point on the side of the house and work down. If the ground is sloping so the ladder cannot be placed against the side of the house without sliding away from an upright position, clamp a 1-by-4-inch board to the leg of the ladder on the side toward which the ladder slides (low side). This makes up the difference in ground level. Use two C-clamps to hold the board firmly in place.

Don't carry a full gallon of paint up the ladder. It is better to work from a half-gallon bucket. An S-shaped metal hook cut from stiff, heavy wire will enable you to hang the bucket from a rung of the ladder convenient to your reach. Then you can hang on to the ladder with one hand and paint with the other.

Be sure to stand the foot of the ladder far enough away from the house, so there's no possibility of its tipping back, away from the house. Never stretch so far away from the center line of the ladder that you lose your balance or cause the ladder to tip.

Do not dip the brush into the paint more than 1½ inch—never to the metal ferrule. When the brush is loaded with paint, wipe one edge after the other against the interior of the rim of the bucket, tip the brush handle down to carry the paint to the siding, and brush the paint on the siding. Brush in all directions; don't overwork it. Make the final strokes light ones along the grain of the siding, with the brush tilted toward the siding surface.

Always allow sufficient time toward the end of your painting session to clean your brushes and other equipment thoroughly.

To prevent the paint remaining in the gallon can from skinning, pour a tablespoon of thinner (water for latex paints, mineral spirits for solvent-thinned paints), over the top to form a thin, continuous film. Then, press the lid tightly in place. The small amount of fluid added over the paint not only prevents skinning, but also offsets loss by evaporation. Mix the paint well before beginning to paint the next morning.

Index

262